The Wisconsin Logging Book 1839-1939

by

Malcolm Rosholt

Malcolm Rosholt

ROSHOLT HOUSE
Box 104
Rosholt, Wis. 54473

Foreword

In the heyday of the logging era, there was a persistent belief among log drivers that logs floating down a river outran the current. It seemed to the drivers that the trees of the forest were not only their formidable opponent, but that they had mystical powers which enabled them to even outdo another of Nature's own forces—the river. For me this myth embodies the essence of the battle in which the loggers were engaged in their struggle to subdue the overwhelming presence of the forest in this land.

My fascination with the logging era of the 19th and early 20th Century is home grown. I was born in a house less than a block from a sawmill, and I grew up with the smell of pine sawdust in my nose. When I was fourteen years old I worked in my grandfather's sawmill, wheeling sawdust in a big wooden wheelbarrow away from the mill and dumping it.

In 1976 I published *A Photo Album of the Past*, a collection of pictures on pioneer life in central Wisconsin, which includes two chapters on logging. Although the two chapters comprise only a quarter of the book, these pictures, most previously unpublished, made such a deep impression on readers that many began calling the *Photo Album* "that logging book." The popularity of the logging pictures has prompted me to do a combined picture and text book on logging covering all of northern and western Wisconsin.

The year 1839 is chosen as a point of beginning because in that year George Stevens of Almond, New York, began building a dam for an "up-and-down" sawmill at "Big Bull Falls" (Wausau). In a letter to one of his financial backers in St. Louis, Missouri, he draws a sketch to show where and how he is going to dam up the mighty falls at "Big Bull." This sketch appears to be the earliest visual evidence of sawmill activity in Wisconsin, although legal documents confirm earlier sites.

The year 1939 is a convenient time to close the book. By that time the chain saw was about ready to take over from the crosscut saw, and trucks were replacing horses. Thus, the book covers, roughly, the first century of logging in Wisconsin.

While the pictures are nearly all from Wisconsin libraries, museums, and private collections, several wet plates, dating to 1864, were found in the archives of the Minnesota Historical Society at St. Paul, and because these pictures seem to predate anything found in Wisconsin on the subject, I crossed the Mississippi River to include this rare collection, confident that the pictures represent the same techniques, tools, and life style of the Wisconsin logger in the 1860s.

I was fortunate to meet and talk with William Hoyer of Cudahy, the last survivor of the Chippewa drive (he always pronounced it "Chippeway"), who died in 1979. Several others whom I borrowed pictures from are also deceased, including Mrs. John Wipperfurth of Tomahawk, William Kaiser of Eau Claire, Roy Martin of Winter, and T.J. Thompson of Bruce.

I am deeply grateful for the art work and diagrams done by Randall E. Rohe, UW-Green Bay, and for the several maps done by students of Ray Specht, UW-Stevens Point. Valuable catalogs and documentary materials were also provided by Gordon Sorenson, Drummond, Howard Peddle, Iron River, and Michael S. Slasinski, Saginaw, Michigan. Archivists who lent generous support were Elizabeth A. Knight, Minnesota Historical Society, Kathy Halsey, UW-Stevens Point, Dorothy Heinrich, UW-Green Bay, Edwin Hill, UW-La Crosse, Robert Walther, the Smithsonian Institution and Beulah Larson, Stevens Point.

On the cover of the book is a picture of a pine stump which stood for more than a century in Iola pond (Waupaca County) before the drawdown in 1965. The giant tree was felled, probably, in the early 1850s. The circle "L" has been superimposed on the stump; it is supposed to represent the stamp hammer mark of Paul Bunyan, water boss on the great Round River drive, in '64 or '65.

Malcolm Rosholt
Rosholt, WI 54473

Library of Congress Catalog Card No.: 80-53389
Fourth Edition Revised 1985
ISBN: 0-910417-05-9

Printed in the United States of America by
Palmer Publications, Inc., Amherst, WI 54406

Contents

One of the last great groves of white pine in private hands in Wisconsin stood north of Park Falls and here are a few, about to be harvested for the Roddis Lumber & Veneer Company of Marshfield in 1926.

Daylight in the swamp

The Pinery of northern Wisconsin takes its name from the great white pines *(Pinus Strobus)* that were found along the banks of the river systems, on high land and low land, in groves, and sometimes as a single tree isolated among hemlock, yellow birch and basswood.

The white pine has become a part of American life and has meant so much to the economy of several states that it has been named the official state tree of both Maine and Michigan, and in Maine the white pine cone with tassel is the official flower of the state.

There is an easy way to identify the white pine: the needles come in bunches of five. In the fall these needles are shucked off by the tree while green ones take their place, and in this manner they seem to grow a shade greener in winter by contrast with branches heavily laden with new snow.

The lumber, laths and shingles from the Wisconsin Pinery gave pioneers in the frontier settlements of the Midwest a chance to move from sod huts and log cabins into frame buildings. In fact, there was scarcely a state in the Midwest that did not build most of its frame dwellings and store buildings from the lumber of this tree, even to the doors, window sashes, mouldings and interior trim.

The white pine is a soft, pliable wood, yet sturdy and easy to work whether with a saw, drawshave, or sculptor's chisel. The pioneers who settled among these pines rocked their children to sleep in cradles made from white pine slats, and they buried their dead in coffins made of white pine boards. No matter how it is viewed, this tree was a precious inheritance from the past.

Today, most of the white pine lumber *(Pinus monticola)* comes from the far West, and the white pine as a timber resource in Wisconsin has been supplanted in large measure by the red or Norway pine because the latter is more resistant to disease and grows faster.

The popular attitude toward trees in general has changed radically in the past century. Time was when the tree was considered a thing to be used for the benefit of the human condition, and if it stood in the way, to cut it down, make lumber of it, or burn it. But in this century, a poem written by Joyce Kilmer in 1913, has done much to change that. Today the tree is an object of adoration, a sublime creation of God, although in many places, especially along some of the heavily traveled highways, the tree is also the cause of many fatal accidents.

Popular belief today holds that the expression "daylight in the swamp" was a morning wake-up call coined in the logging woods of Wisconsin. Actually, it was an expression that originated in Maine and had nothing to do with waking anyone. It refers to the time when the pioneers of Maine were cutting down the trees in the lowlands and swamps and, when a clearing was made, people said they had found, or could see, "daylight in the swamp." J.S. Springer, author of an article called "Joys and Perils of Lumbering" in *Harpers New Monthly Magazine* (March 1851) refers to "swamp life" to describe the work that was going on in the woods, presumably the Maine woods. Getting daylight into the swamp, therefore, was like making an opening, or driving a wedge into the forest, whether in Maine or Wisconsin.

There is a story, probably true, about an early settler in Portage County who agreed to buy a forty of land, heavily wooded to pine timber, from Conlee Brothers of Oshkosh on the condition that the lumber company cut the timber off at once. The settler did this because he was not equipped to cut the timber himself, and he wanted to get it out of the way so that he could begin farming, even if he had to do so between two stumps.

Conrad Richter calls the first volume of his trilogy on pioneer life in Ohio *The Trees* (1940). In it he describes how the pioneers feared the forest and, as children, developed a hostility towards it. In the first sentence of his novel he suggests the gloom or, as he calls it, "the midway twilight of the forest."

In *The Fields*, the second of his novels on poineer life in Ohio, Richter tells what Sayward, the protagonist, is challenged by as she makes her way home through the forest:

> By day the woods seemed more open . . . You could look up and see through hole after hole in the branches overhead . . . But at sundown the woods thickened . . . All around, you could feel the woods

James Parfitt (Tigerton) took this picture of the forest as it looked in the southeast corner of Marathon County, about 1910. It is believed that the picture was taken on the "tote road" that ran from Galloway west to the one-time sawmill town of "Holt."

swarming and crowding, butt [tree] to butt, with branches matted and braided, all shrouded with moss, older than the wild bull's tails and dark as midnight, running on and on a slew of miles you couldn't count.

A century ago, whether in the United States or Canada, the forest was an awesome thing, filled with danger and even supernatural evils like trolls. Roads were narrow, hemmed in by trees that menaced and, in some places, shut out the daylight. A trip to the village through this wilderness was something Father might not be afraid of, but for Mother and the children there were all kinds of possibilities for disaster. In short, the forest was a threat to the white man. It dominated him, and this he could not tolerate. It had to be contained and, if necessary, cut down.

The attitude of hostility to the tree among the Yankee migrants to the Midwest and among the European immigrants who came after them, was similar to the attitude held by the pioneers of Lower Canada. Nowhere is this better illustrated than in a poem which appeared in *Winter Studies and Summer Rambles*, a book published in 1838, and written by Mrs. Anna Brownell Jameson, an English-born poet who came to Canada with her husband in government service. This is what she wrote in part:

A Canadian settler hates a tree
regards it as his natural enemy,
is something to be destroyed, eradicated,
annihilated by all and any means.

In Ohio, the pioneers were so desperate to begin farming that they did not take the time to cut the trees down. They choked them to death by "girdling," that is, they stripped the bark around the trunk to stop the flow of sap. A tree girdled in the fall would be dead by spring, and without leaves, there was enough sunlight for crops to grow beneath it. When the farmer had nothing else to do in winter, he cut the trees down, rolled them into a pile and burned them, or cut them up for firewood.

Clearing bees were organized in Wisconsin to cut trees. Neighbors got together to help a "newcomer" clear a piece of land. Small trees, whether pine or hardwood, were cut down and rolled into piles and burned. Anything to get "daylight in the swamp."

The vast majority of pioneers held no resentment against the lumber companies for cutting down the timber, pine or hardwood. Men like John F. Dietz at Cameron Dam resented the lumber companies not for the timber they were cutting but for the fact that they were not getting a share of the proceeds. The lumbermen were engaged in a process which, aside from a handful conservation-minded individuals, was heartily approved of. A minority of these lumbermen got rich, but in the early days of lumbering, most went bankrupt as a result of inexperience, or lack of capital to tide them over a disastrous fire. By the 1880s, many of the small mill operators had been eliminated and replaced by big stock companies capable of taking their losses with their gains.

Around the turn of the 20th century, anyone who got very rich in the lumber business in Wisconsin came to be called by the muckrakers a "lumber baron," a slur phrase to suggest that he had gotten rich at the expense of others, or at the expense of nature, or by illegal means, overlooking what these "barons" meant to the economy of Wisconsin or to employment. That there were unscrupulous men in the lumber business cannot be denied, any more than it can be denied that these same types are still prominent in both small business and big business.

But most of the men who got rich in Wisconsin, stayed in Wisconsin, built beautiful homes, donated to charities, established foundations, hospitals, libraries and scholarships. One interesting trust fund was established for "special education projects" by the will of Thomas Brittingham, a "lumber baron" practically unknown even to his contemporaries. In 1979 the interest from the $25,000 trust fund he had established was used to pay the expenses of a delegation of faculty members from the University of Wisconsin-Madison to the People's Republic of China.

Far more important, the money accumulated and the experience gained by these men in the forest industry led directly to the manufacture of paper products and the building of paper mills such as Consolidated at Wisconsin Rapids, Kimberly

Clark at Kimberly, Fort Howard at Green Bay, and others in Wausau, Brokaw, Rhinelander, Neenah-Menasha, Nekoosa and Kaukauna.

It is sometimes held that the lumbermen cut down the "virgin pine" and left the slashing, following a policy of "cut out and get out." But there never was any virgin pine. A forest grows only so high, and by a process of natural selection, it is constantly dying and constantly renewing itself. The only exception to this might be the great redwoods of California.

That trees need to be thinned, is a foregone conclusion. In 1979 a county forester made a boring on a tree growing in the author's tree plantation. The tree, a species of aspen, was nearly two feet in diameter. The forester counted eighty-seven rings, but, he said, the tree had stopped growing twenty-five years ago.

There was no forest crop law in Wisconsin until 1927, and selective logging before that time was impractical. Timber was taxed on its value each year, and since most resources increase in value with an expanding economy, the taxes on the timber went up each year, adding further incentive for the lumbermen to "cut out and get out."

Where the lumbermen of the past failed was in their disregard of reforestation. When they bought a forty of timber land, they seemed to think it was their privilege to cut down every last pine tree no matter how small. For example, in the winter of 1911-12, Brooks & Ross of Schofield logged nine forties of scattered "swamp pine" southwest of Elderon (Marathon County), many of the logs scarcely big enough to make a two-by-four studding. This also precluded reseeding by nature.

Meanwhile, the lumber camps and sawmills gave employment to thousands of unskilled and semi-skilled workers, many of whom had emigrated from the "old country" and were willing to work from daylight to dark, often in the bitter cold, if only to hold a job and be assured of three "square" meals a day.

The lumbermen started small towns and attracted businessmen and stores, blacksmith shops, feed mills, hotels, restaurants, and saloons! Roads were built between these towns, which opened up communications, and the railroads brought in supplies and took out the finished lumber.

Corner of a log cabin probably built by Jens Mortensen, an immigrant from Denmark who settled in Alban township (Portage County) in the early 1860s. The building technique used here is called "dovetail notch," a style brought to the United States from Sweden in the 17th century. This picture was taken in 1957 but the building has since been demolished.

In an act passed by the state legislature in 1898 entitled "Of Liens," a fair idea is given of the scope covered by the lumbering industry in Wisconsin in the latter half of the 19th century. The act was passed, no doubt, to protect the rights of workmen who had performed any labor or service "in cutting, hauling, running, felling, piling, driving, rafting, booming, cribbing, towing, sawing, peeling or manufacturing into lumber or timber any logs, timber, stave-bolts, staves, pulpwood, cord wood, railroad ties, piling, telegraph poles, telephone poles, fence posts, paving timber, tan or other barks or in preparing wood for or manufacturing charcoal, etc."

To this might be added all the factories that were built in Wisconsin to manufacture sash for windows and doors, blinds, pike poles, rafting wedges (grubstakes), furniture, shingles and all kinds of wooden ware such as staves, tubs, and pails, the largest of these being the Menasha Wooden Ware Company which had mills and branches throughout the state.

The "logging days," as they are usually referred to in the early period of Wisconsin history, were wasteful of wood and even of life, but it was in the nature of the times to be wasteful of both. For example, opening up the forests to railroad traffic in the 1870s opened the way to fires caused by sparks from the engine. Farmers also started brush fires that got out of control, and lightning caused fires that consumed thousands of acres.

Despite the disastrous fires such as those at Peshtigo in 1871, and in the northwestern counties in the early 1890s, including the great Phillips fire, the nation and the state made little effort to counter forest fires with anything faintly resembling a forest service, for just what could be done, at the time, to stop a fire? Very little. Horse-drawn wagons with local crews of loggers were rushed to the fire zone to dig ditches or set counter-fires, but beyond that there was little they could do except wait for nature to bring rain or turn the wind. Almost every big sawmill had a number of water barrels on the ridge board of the roof, but there is no record that any of these carefully poised water barrels ever doused a single fire. Mills burned like match boxes and there was little time for anyone to crawl onto a roof and spill water on the shingles.

Fire in the timber was always a serious matter, not only as a threat to life, but because of the losses to commercial timber that either burned or blackened. It could affect small operators as well. For example, a farmer north of Rosholt (Portage County) had a small crew working during the winter cutting railway ties which were decked at a siding on the railroad. In the spring, a fire started in the grass nearby and ran down the tracks to wipe out the work of an entire winter. The farmer had to mortgage his farm to pay the crew.

But anyone who wishes for a return to the so-called delights of the "forest primeval" is drawn by an illusion which, like the Golden Age, never existed.

With the pine forests nearing exhaustion, lumbermen began to make paper from wood products unsuited to the manufacture of lumber, such as aspen, and spruce, and soon demand outgrew the supply, and paper mills were expanded into giant corporations which today employ many thousands more men—and women—than ever worked in the logging woods or in the sawmills. Pulp for the paper mills is also coming to Wisconsin from Canada, but much of it still comes from local wood lots scattered over the state. One crop of aspen renews itself about every twenty years and now there is talk of growing trees in six years!

Some companies, such as Connor Forest Industries, and Tigerton Lumber Company, have acquired sufficient timber holdings in Wisconsin and Upper Michigan to do selective cutting in both hardwoods and soft, thus assuring themselves of a perpetual supply of timber. The paper mills on the Wisconsin River also have tree plantations scattered throughout the north to back up their supply of pulp wood.

While the tree is still an object of adoration, and always will be, it would be difficult to convince any of the dairy farmers in Oconto or Marathon counties, or the potato growers in Portage and Langlade counties that their land should be replanted to white pine. The beautiful farms, some with four and five silos, that dot the land where the white pine once grew, probably produce more food and more wealth on one acre than the tallest white pines on forty acres because grain and vegetable crops spring up annually, the pine tree only once in fifty years.

Still, the beauty of Wisconsin lies in its remaining trees that grow on land unsuited to cultivation, in its lakes and rivers. In fact, one of the residual benefits of the lumbering era is the many ponds, lakes and flowages left behind by sawmill owners who dammed up creeks and the big flowages originally for the purpose of log

Picture reproduced from commercial postcard titled "Road to Holt from Galloway, WI." Parfitt took the picture, about 1910.

driving such as on the Flambeau, the Wheeler, and the Gila. In some cases, dams have been restored to one-time driving streams such as at Post Lake on the Wolf, and Lake Winter on Brunet River. The earth belongs to the living, and what the lumbermen did was, under the circumstances in which they lived, probably inevitable. It seems appropriate to quote Professor Charles E. Twining, author of *Downriver: Orrin H. Ingram and the Empire Lumber Company* (1975):

> In years to come much would be written about the destruction of the great forest; and perhaps too much would be forgotten about the processes involved, and the cheap lumber produced, and where it went, and what it did for an aggressive, westward-expanding America.

It is time now to turn the leaf of history and to have a look at the "logging days," to learn how the logs were cut, hauled and floated down the rivers of Wisconsin. This great epoch in Wisconsin's past was a period of testing and of hard work, but the majority of the men who labored through it, liked it and came back to camp year after year, feeling lost without a mess hall filled with hungry men, or a sleeping shanty reeking with odors of wet socks, or the sound of French-Canadian love songs, or the suspense of a good story.

In the woods and "on the haul"

The basic work unit in the logging woods of Wisconsin around the turn of the 20th century included four men: two sawyers, a swamper, and a skidding teamster. Earlier, when oxen were used for skidding, a fifth man, called the "chainer," was required. After a log had been rolled onto a travois or go-devil, it was the chainer's job to secure the chain holding the log to the travois, and to run the grab hook forward to the chain that went, between the oxen, to the ring under the wooden yoke borne over the shoulders of the oxen. Most of the time two or three yoke of oxen were used for skidding. A big log could require four.

The swamper limbed, that is, cut off the limbs from the tree felled by the sawyers, and he cut away brush and small trees that might stand in the way of the teamster. Usually, one good swamper could keep up with the teamster, but sometimes the underbrush was so thick, or windfalls so numerous, that two swampers were needed.

It took two sawyers to handle a crosscut saw big enough to fell a tree and "buck it up" into suitable lengths, usually sixteen feet. (Old loggers from Maine to Wisconsin used the term "to fall" a tree, not "to fell" a tree.) These measurements were imposed by the length of the average sawmill carriage, and the length of the average logging

Two sawyers "bucking up" a log into standard lengths of eight, ten, twelve, fourteen and sixteen feet. The picture was taken in 1864 in Elias Moses camp on Tibbett's Brook, a branch of Rum River, northwest of Milaca, Minnesota. Joel E. Whitney operated the "Sky Light Dagerrean Gallery" in St. Paul at this time and either Whitney or an associate, Charles A. Zimmerman, took the picture. The wet plate is not reproduced from a tintype, but was shot with a camera that made a frame about eight and a half centimeters wide and eight high. The two sawyers are using a crosscut saw. It is a warm day in the woods; the sawyers are working in shirt sleeves, and neither has mittens, although high leather boots were still in. Behind the sawyer at the left stands a "chopper" who is attacking another tree.

sleigh which one or two teams of horses could pull without straining themselves.

It was a matter of pride for the sawyers to fell a tree in such a way that it did not damage other trees, or the tree being cut, and to prevent the falling tree from getting caught in the branches of other trees. A tree poised in this manner was called a "widow maker" because it was apt to fall when least expected. Usually the only way to remove a tree that is hung up or leaning on another is to cut down the tree blocking the way, or to fell a second tree against the first one, in either case a risky business with a strong chance of damaging one or both the trees.

Four years after the picture on the opposite page was taken, it appeared an an engraving in *Harper's New Monthly Magazine* for March 1868 in an article titled "The Minnesota Pineries." But the chopper in this engraving is badly off balance. Since the chopper was right-handed, his left foot should have been forward, not his right. It also seems that the tree he is felling is going to get caught in the tree at the left, and that will create a "widow-maker."

Before the sawyers were to fell a tree, one of them notched it with an axe on one side, and applied the saw from the opposite side of the tree, at an angle, slightly above the level of the notch, in such a way that when the tree fell, it not only fell clear of other trees, but it did not leave a big splinter behind on the stump, often called a "barber chair." The depth of the notch chopped into the tree depended on its size, and the side the notch was made on determined the direction in which the tree would fall.

Before the woodsmen went into the woods in the fall, the timber had been inspected by the "cruiser," a man paid to estimate how much lumber a given forty contained. To mark off a forty of timber, the cruiser often blazed or cut a piece of bark on the trees lying on the quarter line, the blaze on the outside indicating the limit of the owner's proproperty. When the sawyers began cutting, the tree was not supposed to fall across the quarter line, otherwise the man who owned the adjoining forty could claim it. But the men in the woods were not very fussy about these matters, and legend has it that it was permissible to cut any tree along the quarter line as far as a man could throw an axe into the next man's forty. In some instances, unscrupulous lumbermen bought one forty of land, logged the timber on it, and then cut from the three forties around it. This was called the "round" or "big" forty.

Once felled and "bucked up," the logs from the trees were skidded out to a tote road by a team of horses or yoke of oxen, and from the tote road the logs were hauled on a sleigh to the nearest skidway, landing, rollway or banking ground for winter storage. In the early period of lumbering in America, roughly 1825 to 1875, oxen were used to skid the logs, and they were also used to haul them to a skidway. After 1875 horses began to replace oxen and by 1910 horses were being used almost exclusively in the woods. The main reason for this was that by 1910 a bigger breed of horse was available for logging purposes. This breeding to a larger horse continued into the 1930s when some weighed more than a ton, usually "Perches" (Percherons), and mixed Percherons and Clydesdales.

Before winter operations in the woods began, a logging road had to be laid into the timber, and as the quest for more trees continued, a network of roads developed. At times like this, care and planning were vital in order to prevent duplication of routes, and also to take advantage of the terrain, to follow the lines of least resistance, and, if possible, avoid any hill however small. Roads used for the sleigh haul to a skidway or to a landing on a river often followed the bank of a river or creek, or traversed a swamp because here the terrain was usually flat.

A road leading into the logging camp from a traveled highway was called a "tote road," and the teamster who hauled in supplies to camp was driving the "tote team."

Most landings, banking grounds, and similar places had multiple skidways. And a skidway was a place where several logs had been laid on the ground parallel to each other to form a base on which the logs were decked. It was called a "skidway of logs," not a pile of logs. It usually took a crew of men two to three weeks to roll the logs from the skidways into the river. This was

The basic work unit in the Wisconsin logging woods was a crew of four men, like the one pictured here, a skidding teamster, a swamper, and two sawyers. When oxen were still being used for skidding, a fifth man, called the "chainer," was part of the crew. It was his duty to run the chain around the log and carry it forward to the ring that was attached to the under part of the oxen yoke. This picture was probably taken in Sawyer County for Kaiser Lumber Company of Eau Claire.

called "sacking."

Before 1851, loading logs on a bob sled in Maine, and across the St. Croix River in New Brunswick, was a primitive affair, and though the principle of "loading cross-haul with single chain" had been developed, it apparently was not being used in the logging woods. In the article written by J.S. Springer, mentioned earlier, he describes a bob sled being loaded "sometimes [with] three large trees . . . " And then, he adds, "all things in readiness, the men follow up with handspikes and long levers."

This suggests that loading logs on a bob sled was an art still in its infancy. All that could be hauled on these sleds was "three large trees" (it can only be assumed that the writer is referring to logs, not trees). It also seems clear that the logs were rolled on the sled by a crew using handspikes and long levers, patently a task for muscle men only. A handspike was merely a hardwood stock, less than two meters long, sharpened to a point at one end. It would be difficult to guess what a "long lever" was unless it was an extended handspike or "jam pike." There was an early type of cant hook called a "cantdog" in use by this time, and it may be that Springer was referring to the cantdog as a "long lever."

Sawing Down Trees.

This illustration appears in the *Lumber And Log Book*, a pocket-sized book for reckoning log measurements, and other tips to lumbermen. The first editions were published in the 1850s, and the illustration shown here appears in the 1880 edition. The engraver, however, was not aware that no one would saw a log this close to the ground. Most pine stumps were at least a meter high. There does not appear to be any notch in the tree either, but it is significant that the illustration shows a single-handled cross-cut saw, not the more common two-handled saw. It also seems to be the earliest evidence that trees were now being sawed down, not chopped down.

These two sawyers were probably working for Kaiser Lumber Company in a camp located east of Winter (Sawyer County) in the early 1900s. Either the swamper or one of the sawyers has trimmed the bark around the tree to make it easier for the cross-cut saw. The short piece of string holds two wedges which have been driven into the kerf to keep the tree from pinching the saw. The wedges are tied to a string to keep them from becoming lost in the snow when not in use. The pine tree the two men are cutting appears to be the last. In the background nothing seems visible but slashing, "rampikes" and underbrush.

Three yoke of oxen skid a big log through the snow, the butt end balanced atop a travois of some kind. The engraving is enlarged from a sketch appearing in "Life Among the Loggers", *Harpers*, March, 1860.

Four yoke of oxen were often used to skid a big log out to a logging road. This is a scene from the camp of Abe Johnson on Bone Lake east of Milltown (Polk County), 1890s.

The pride of youth in doing a man's job is reflected in the face of this young ox teamster somewhere in the Chippewa River valley. The oxen seem perfectly matched. The yokes over their shoulders do not appear to be as heavy as some noted in local museums in the state. The teamster holds a "goad" stick which he uses, not to strike the oxen, but to tap them on the cheek or head to indicate which way they are to turn.

The original photograph, taken on January 20, 1892, is captioned, "Four yoke over a hard spot, Brewster's Camp." The oxen are pulling on a load of logs (left) and Brewster was probably jobbing for Knapp, Stout & Company north of Menomonie on the Red Cedar River.

This is a small version of what was once called a "barber chair". The tree has not been sawed properly and it has broken off, leaving a big splinter.

A small travois, or go-devil, used for skidding logs out of the woods, this one was made from a natural deformity in a tree. For skidding, a log was rolled onto the crosspiece, and a chain, wrapped around the crosspiece at the left, went over the log and under the right side of the crosspiece and then forward to the iron bar in front. After a loop around the iron bar, the chain was hooked to the clevis on the evener behind the horses.

Teamster Otto Newbauer skids out a log with chain hooked to skidding tong for Tigerton Lumber Company.

Chain loading

Just when the technique of loading a load of logs with single chain, often called "chain loading," was introduced into the logging woods is not clear, but from the evidence of an article written by J.S. Springer for *Harpers*, it was not before 1851. The technique may be explained this way: two saplings, peeled and pointed at one end, called "skid poles," are laid against a sleigh or wagon to serve as a ramp on which the log is to be rolled upward. A log is then canted to the side of the sleigh (or wagon) and a chain slipped around the middle of the log. With one end of the chain hooked to the sleigh bunk, or to logs already on the sleigh, the other end of the chain is hooked to the evener behind the horses (or oxen) on the opposite side of the load from the log, and when the horses pull *away* from the load, the log rolls up the two skid poles like a wheel being pulled up an incline.

An illustration of this technique appears in the *Lumber and Log Book* (1880), a small handbook for lumbermen on such techniques as scaling lumber and other tips for successful lumbering. The caption under the illustration in the *Log Book* reads: "Loading Logs on a Wagon—The Cut Explains Itself." But the cut does not explain itself entirely, for it would seem unlikely that anyone would roll a log up two skid poles without a man to guide them, the so-called "send up" man, otherwise the log will probably swing to one side or the other and land askew on top of the load.

One of the pictures taken by a cameraman for the Whitney Gallery of St. Paul in 1864, depicts a crew loading with a team of oxen on "cross-haul," probably the earliest pictorial evidence in the Lake States of this technique. In Wisconsin, the cross-haul was used for loading both sleighs and railroad cars before and after the turn of the 20th century. (A picture taken in the logging woods north of Rosholt [pages 28-29] in the winter of 1903-04 shows a crew loading flat cars with single chain technique.)

The picture taken for the Whitney Gallery in 1864, which is duplicated in an engraving in an article entitled "The Minnesota Pineries" in *Harpers* (March 1868), does not show a man on top of the load, the so-called "top loader" who guided the logs into place, one tier above the other, careful to note when a chain, called a "wrapper," was needed to bind the logs together as the tiers built up, higher and higher.

The top loader had one of the most dangerous jobs in the woods, but there were seldom any openings for his position. He was a man quick on his feet, with a voice that could be heard far and wide, and when he shouted for the teamster to "whoa" he meant "whoa" not "gee" or "haw." Some horses became so well-trained that they did not require a teamster, but were guided by the commands of the top loader.

In Michigan an important advance in loading

operations was made with the introduction of the steam engine called the "donkey." This had a boiler that stood upright on a platform, and could be moved about, slid back and forth under its own power, and had more horse power than there were horses in camp. Patented in 1882 by John Dobeer, it consisted mainly of a small, high pressure steam engine which could be used to turn, through a set of gears, a capstan like spool or drum to pay out rope or cable like a line on a fisherman's reel.

By 1886, Horace Butters of Ludington, Michigan, developed a new skidding technique with the donkey. Instead of skidding the logs on

Loading Logs on a Wagon—The Cut Explains Itself.

This illustration appears in the *Lumber and Log Book* (1880) and is reproduced here actual size. The engraving attempts to explain the technique of loading cross-haul with single chain. The principle is clear: one end of the chain is hooked to the bottom of the wagon in this instance, and the chain looped around the log and then back to the oxen on the opposite side of the wagon. But there is no "send-up" man to "cut" the log, and the log will almost surely go askew before it goes much farther.

Another wet plate by the Whitney Gallery taken on or near Rum River in 1864. This one provides what is probably the first visual evidence of loading logs on a sleigh by "cross-haul with single chain." The oxen are pulling on a chain that is rolling the log up the skid poles where two men are guiding the log. There is no top loader, probably because the size of the load does not warrant one. The logs rest directly on the sleigh beam, not on a "bunk," (or "rocker" as it was called in Minnesota). There is also evidence of a "spring pole" atop the load, apparently a device to tighten the chain holding the second tier of logs. This wet plate, slightly modified, also appears as an engraving in *Harpers* for March 1868.

the ground to a landing for loading, he stretched a cable, anywhere from three hundred to six hundred meters long, between two tall trees, their limbs cut off to make a pair of spars, one called the "head spar," the other the "tail spar." A trolley traveled on this cable and when a line was lowered from it to the ground to pick up a log, the log was lifted clear of the underbrush and whisked back to where the loading operations were in progress. It was called "high lead" logging.

Oral tradition holds that the Hines Lumber Company of Chicago tried this method in northwestern Wisconsin, but it was not successful. The drawback was that it took just as much time to lift a small log as a big one out of the slashing. But the technique found favor in the Pacific Northwest where mostly large trees were harvested. The technique used in these operations is suggested by a photograph of a sketch found in the iconographic collection of the Minnesota Historical Society, but the scene seems to be related to the Pacific Northwest, not Wisconsin or Minnesota.

Except for Hines, it appears that Wisconsin lumbermen were satisfied to use horses, or oxen, to skid their logs out of the woods although here too there may have been exceptions. A photograph taken of a "steam log skidder" on the Stanley, Merrill & Phillips railroad, about 1915, suggests that logs were skidded out of the woods to the railway on a long cable, probably along a trench or trough made of split logs. The picture shows a jammer boom on either end of the flat car supporting the cab and engine works. This method of skidding logs out may also have been used first in Michigan. A sluice or beveled V-shaped trench, made of logs hewed on one side, was laid out through the woods, and a cable was run out from the donkey engine to the deck where the logs were waiting to be picked up. It was, in effect, a dry flume which, instead of using water to float a log, used a cable to skid the log along a trench.

Before 1900, the Rust-Owen Lumber Company of Drummond used a mounted trough, or flume, filled with water, to transport their logs from the hill country west of Drummond to their mill in the village. But this is the only instance presently known of a Wisconsin company using a flume in

a manner that became common in the mountains of the Pacific Northwest.

These devices are mentioned to suggest how many ways there were of getting a log out of the woods, but it also seems clear that, while some of these innovations worked, others did not work well enough to warrant the time and expense to install them.

The donkey engine found other uses in Wisconsin. A picture taken in 1892 near Menomonie for Knapp, Stout & Company shows a donkey engine pulling a load of logs by long cable over a hill.

The technique of loading logs "cross-haul with single chain" is well illustrated in this picture taken at August Selmer's camp on Comet Creek (Waupaca County) February 2, 1911. Selmer stands second from the right and his son, Roy, is top-loader. The chain has been looped around the center of the log and the horses at the right will pull on the chain, causing the log to roll up the skid-poles to where the top-loader will decide where it will fit on the load. Most logs were heavier, or thicker at one end. The chain had to be looped slightly towards the heavy end to achieve balance. But the man holding the cant hook in this picture, called the "send-up" man, or "sender-upper," had the responsibility of seeing that the log did not become unbalanced in its progress up the skid-poles. If he saw that the log needed correcting, he "cut" it, that is, he adjusted it with his cant hook. The expression "to cut" a log was heard in both Wisconsin and Minnesota. The chain rolling the log in its upward ascent was disengaged when the log rolled over the pup hook attached to the end of the chain.

About 1890, the engine was mounted on a flat car, a shack built over it and it was used to load logs on flat cars. This type of "steam jammer" probably was patterned after the steam-powered derricks used in railroad building to pick up rails and ties in front of track-laying crews. But this combination of donkey engine and jammer was limited in its capability for loading logs on flat cars. The problem was: how to get the flatcars around the jammer without putting in a railroad spur, or without running the jammer and flat car to a switch on a siding or to a "Y."

"Ginpole Johnson" and the woods jammer

Some time in the early 1890s, according to oral tradition, a portable jammer was developed by Charles Johnson, an immigrant from Sweden to Munising, Michigan. Johnson became known to his contemporaries as "Ginpole Johnson" because he employed the ginpole principle in developing his jammer for log loading. But whether developed by Johnson or someone else, the portable woods jammer was being used in Minnesota in the early 1890s, and a picture taken at Gordon, Wisconsin, in 1903, shows an A-frame jammer in operation.

Charles ("Ginpole") Johnson, an immigrant from **Sweden** to Munising, Michigan, has been credited with developing the woods jammer in the early 1890s.

Eight former lumberjacks, all in their eighties when interviewed by the author some years ago, agreed, independently, that they had seen their first woods jammer some time between 1906 and 1908, either at Big Falls in Waupaca County, Galloway and Mosinee in Marathon County, or at Kempster in Langlade County.

There are variations of the portable jammer. One was the ginpole which had only one pole reaching either straight up through the trees, or at an angle. Another was called the "A-frame" with two poles meeting at the top like the letter "A;" and the third type had a boom swinging out from the ginpole called the "swing-boom" jammer.

No mechanical power was needed. Instead, a team of horses was used on cross-haul. A cable ran from the evener behind the horses to a toe block at the base of the ginpole, thence up to a block on top of the pole, and from there down on a traveling block to a crotch chain. After a "pup hook," had been hooked to either end of the log, the log was hoisted into the air by the horses pulling on the cable. The pup hookers each held a long rope, called the "sucker line", attached to the pup hooks, and, at a signal from the top loader, the lines were yanked and the hooks were disengaged, releasing the log.

Although it was most commonly called the "woods jammer," it was also known as a "side jammer," probably because it was used extensively for loading flat cars and was usually stationed on the side of the cars. The A-frame jammer continued to be used to the very end of the logging era before machines replaced horses. Tigerton Lumber Company was using an A-frame powered by a Ford tractor into the 1950s, at which time the hydraulic lift began to replace all jammers, large and small.

The construction of a jammer was simple. It was made almost entirely of logs to the specifications of the camp blacksmith. The upright logs or booms rested on two chamfered skids which could be pulled in either direction by a team of horses. Had it been developed fifty years earlier, it would have saved a great deal of back-breaking labor. By 1915, at least, a swing-boom jammer was being manufactured by the National Iron Company of Duluth, and it is listed in their catalog for that year as the "Forest Log Loader and Decker."

No one seems to be able to suggest where the word "jammer" came from. Most dictionaries, then and since, have ignored it.

The two photographs, right and below, were taken near Gordon (Douglas County) in 1903, and show the use of the jammer for loading operations both in the woods (right) and on a railway siding (below). The pictures are also the earliest evidence presently available of a jinpole, or woods jammer in operation in Wisconsin. The logs in the picture at the right appear to be suspended in mid air, over the sleigh bunks, with only the face log held by a crotch chain. The picture is not sharp enough to determine the reason for this illusion. The picture below shows a big log about to be rolled up the skid poles, a log especially arranged for the benefit of the cameraman, since the other logs on the flat car run small. This picture provides what appears to be the earliest evidence in Wisconsin of the use of car stakes on flat cars for logging purposes.

Left: Camp 7
Belden Lumber Compa
near Bruce (Rusk Count
some time before 19
Shows gin pole jamm
with slings. Whenever
logs were not large on
average, several w
wrapped together and
chains hooked to
crotch chain for the lift
the top. The top loa
(left) guides the logs
they approach. The t
men on the grou
holding "sucker lin
yank the lines at
command of the t
loader. The jammer
secured by guy wires
cables. Work is proceed
so rapidly in this sce
that the horses have
been unhitched.

A swing boom jammer, this one was called the "Forest Loader." It was a manufactured product, not the work of the local blacksmith, and it had a clutch which made it easier to handle the chain. The cross haul team is not visible. The picture was probably taken in Bayfield County.

Left, two teams of horses were needed to pull this load of logs off Deer Lake (Sawyer County) in 1928. Seth Nichols, a teamster, stands at the left.

"Decking Logs, Northern Wisconsin" is the caption superimposed on this picture. Instead of a single chain on cross-haul, this crew is using a double chain to roll the logs up the skid poles. The jammer, not seen in the picture, stands at the left. Some of the logs in the pile at the upper right are hollow. This kind of timber was cut, lumberjacks said, by the "Hollow Butt Lumber Company."

In the two winters of 1903 and 1904-5, Brooks & Ross Lumber Company logged off the white pine standing north of Rosholt village (Portage County). A spur was extended from the Chicago & North Western line west a short distance into the timber belt, and here loading operations are under way, using the cross-haul technique with the team at the right. The logs were shipped to the company sawmill at Schofield.

The "hoisting machine"

On January 7, 1889, an advertisement appears in the American *Lumberman* for the "New Decker Log Loader" (named for Geo. W. Decker of Newport, Arkansas). It is billed as the "only locomotive log loader," and travelled five miles an hour under its own power, did its own switching, and loaded 100 to 250,000 feet of timber on cars daily with a crew of four men. Within a few months engineers at the P.N. & F.M. Roots Company in Connersville, Indiana, had taken over the manufacture of the "Decker Log Loader" and began advertising in the *Lumberman* on March 9, 1901. This loader solved the problem of how to move a flat car around a loader by building four big cleats, or "spuds," which were lowered to the railway ties when the loader was in operation. When it was time to move, the wheels, or trucks, which meanwhile had been tucked up under the cab of the engine works, were lowered to the railway tracks. This in turn raised the cleats off the ground to make the unit mobile. More important, the engine works stood high enough off the ground for flat cars to pass *under the engine works and through the cleats* in either direction, thus avoiding the need to switch cars to a siding or a "Y."

Two years later, a fairly comprehensive article on the log decker, datelined Cadillac, Michigan, appeared in the *Lumberman* (March 7, 1903) accompanied by a number of photographs. The author of the article refers to the loader as the "McGiffert," but he gives no clue to the derivation of this name. Actually, this loader was manufactured by the Clyde Iron Works of Duluth, a company founded by C.A. Luster in 1899 under the name of Northwestern Manufacturing Company. The name "Clyde Iron Works" was adopted in 1901, and in 1902, the first loader was turned out by the company called the "McGiffert." This was an instant success, and the company could scarcely keep up with demand.

Within a short time, at least two other steam loaders were being manufactured in the Lake States. One of these was called the "Cody" or "American Log Loader," which was manufactured by the American Hoist Company of St. Paul, a company organized in 1882; the other loader was called the "Raymond" which was made by its inventor, O.L. Raymond of Marinette, and advertised in the *Lumberman* in 1908. The Cody resembled the McGiffert in one respect: flat cars could pass under it, but the machine itself did not rest on cleats when loading logs, but on a special flatbed riding on railway trucks over which the flat cars passed in order to reach the other side of the loader.

The Raymond Jammer was merely a steam donkey and winch enclosed in a shack with a boom extending outward, and the whole resting on two skid poles atop a railway flat car. With the skids underneath, the machine could be pulled along the top of a flat car in either direction by running a cable to one end of the car and winding up the cable. It was commonly known in the woods as "slide-ass" jammer.

One of the earliest jammer pictures in central Wisconsin was taken by James Parfitt near Galloway in 1908. Parfitt refers to it in a picture postcard he distributed, as a "hoisting machine," which is true, but it was actually a McGiffert loader.

The inscribed caption on th picture, enlarged from a postca published by Parfitt read "Hoisting Machine" at Gallowa 1908, but it was probably taken c the spur that ran from the Nort Western line at Galloway west the sawmill town of "Holt." Parfi apparently was unaware that th "hoisting machine" was a McGiffe loader manufactured in Dulut This particular loader was used fo more than a decade aroun Galloway-Elderon and was dubbe "Sport." →

A close-up of McGiffert loader Fosterville, once situate near Presque Isle (Vila County). The cleats c "spuds" on the loade are down, the whe trucks up, allowing a empty flatcar to pa under and through th loader. The two line held by the men a "sucker lines" whic were attached to th "pup hooks" on eith end of the crotch chai and when pulled har released the lo suspended in air. Th cars are being loade V-shape, calle "Canadian peakers. The flat cars appear t be "Russel" cars, tha is, a car without a be but with beams an bunks on the whe trucks. The cars wer made by the Russel Ca Company in Detroit i the 1890s, and wer about twenty-feet lon and carried only on deck of logs per ca The initials "V.C.L CO." represent Vila County Lumbe Company. A. (fo Arthur) J. Kingsbury who operated a studi in Antigo, took th picture. It was made i a commercial postcar and copied and sold b the H.N. Richardso Studio in Stevens Point

HOISTING-
MACHINE.
GALLOWAY, WIS.
No 2
1908

V.C.L.Co. 47
Kingsbury
Antigo.
A Log Landing Scene,

The Clyde Iron Works built a standard and narrow gauge jammer. The one seen here is loading pine on a Russel car near Delta (Bayfield County) in 1911 for the Edward Hines Lumber Company. The logs were being hauled to the Hines mill in Iron River over trackage of the Duluth, South Shore & Atlantic railroad.

A steam loader has stopped while the crew makes repairs to the crotch chain (right) and also to the jammer boom. The engine appears to be a "Lima Shay" manufactured in Lima, Ohio, but the picture was taken at Phelps (Forest County).

A flat car has been derailed at Grandview, (Bayfield County), on the Chicago, Minneapolis, St. Paul & Omaha railroad, and two steam jammers are working together to hoist a derailed flat car. At the left is a Raymond jammer, and at the right, a McGiffert. The Raymond was mounted on a flat car and could pull itself on skid poles along the tops of the cars, powered by an upright donkey steam engine. The McGiffert was mounted on wheels and could travel on its own power.

E.H. Hadley, who took this picture, was not sure about the spelling of "skider," but he is referring to the steam log skidder which is being used to skid the logs from the woods on a long cable back to the railway where they were being loaded directly on to the flat cars spotted along the Stanley, Merrill & Phillips line, about 1915. The steam skidder was not widely used in Wisconsin and explains why there is a dearth of pictures relating to it.

A donkey steam engine and an A-frame jammer were used in these loading operations at Day Siding, located half way between Galloway and Elderon on the Chicago & North Western line for Hatten Lumber Company, 1915.

P.N. Burg, owner of a ge store at Shell Lake (Wash County) liked this pictur well he used it on a calendar put a caption on it "Sun Logging Early Days at Lake." The picture was act taken near Bear Lake in and shows the "high whe used for skidding logs out o woods to a nearby skidway the Shell Lake Lun Company, an affiliate of Weyerhaeuser combine. In background (right), oxen being used to skid the logs t high wheeler. The ti harvested here was haulec sleigh and railroad to company sawmill on Shell I after which the village too name. From 1888 to 1901 mill was sawing four million of lumber annually. —

Robert Ball did this sketch of a logging operation presumably based in the Pacific Northwest. But the technique of lifting a log, or logs, out of the forest on long cables with a traveling block was first developed in Michigan in the 1880s.

A Cody jammer, probably in Iron County, about 1910. The Cody was the rival of the McGiffert, but instead of lowering the wheels to travel on its own power like the McGiffert, the Cody was built directly on to the wheel trucks of the railway within a steel framework that supported the engine house. But, like the McGiffert, empty flat cars could pass under the steel frame and be spotted ahead of the jammer as needed. In this picture the jammer seems to be under repair as witnessed by the ladder running forward from the platform. This would have been in the way when loading operations were in progress. The Cody was made by the St. Paul Hoist Company.

Birth of the cant hook and peavey

While handspikes had been used since time immemorial to move a log, one of the earliest mechanical devices was the "ring dog" which was an iron ring around the center of a wooden stock about a meter long with a dog or spur welded to the ring. It took two men to handle it and, no doubt, it was reserved mostly for large logs. It could not be used effectively to cant a log, only to move it slowly to a wagon, sleigh, or saw carriage.

The ring dog was probably followed in Europe by a "cant dog" of some kind—that is, a dog suspended on a stock. This may have originated in one of the German principalities where it was known as *kanthaken*, and in France as *Renard* (fox). It appears in Diderot's *Encyclopedia* of 1768.

The early cant hooks in Maine had a dog hung by a staple or ring bolt on a stock about two meters long, according to Caleb W. Scribner, and the late Lore A. Rogers, founders of the Lumberman's Museum at Patten, Maine. Scribner and Rogers believe it came to Maine by way of French-Canada and was called a "swing-dingle" or "swinging-bitch," a name which, they said, probably "reflected the low opinion in which this inefficient tool was held . . . ."

To roll or cant a log with a swing-dingle, it was necessary to hold the dog in place with one hand

The most primitive tool for moving a log was called the "ring dog," displayed here by Miles Turttelot at the Menominee Logging Camp in Keshena. In his left hand he balances a "jam pike," also a primitive tool which was used on the river drives before the introduction of the peavey. This particular jam-pike was recovered from the bottom of a river in Wisconsin, still fairly well preserved.

A practical cant hook was two hundred years in the making. First came the *kanthaken*, a tool used in the German principalities in the 18th century; the ring dog cant hook and swinging dingle followed the *kanthaken*, and the eye-bolt cant hook probably followed the swinging dingle before the cant hook with straps to hold the pivot eye was used. All these types were replaced in the 1890s by the modern cant hook which has not changed much since that time. At right, a close-up (top) of a *kanthaken;* (center) a high lift cant hook; and (bottom) a ring dog cant hook, all in the private collection of Mike Slasinski, Saginaw, MI.

and insert the spur or steel point into the bark of the log, while the other hand held the stock. It was actually a menace on the river drives as well as on the landing because a log thus engaged could become unmanageable if it began to turn too fast, or roll over in the water. A logger trying to save his swing-dingle under these circumstances was apt to land in the water before he could disengage the spur from the log.

It seems reasonable to assume that the English *cant hook* is derived from the German *kanthaken.* The word *cant hook* appears in the *New England Dictionary* in 1848, although the word most commonly used in the logging woods, at least in the Lakes states before 1900, was "cant dog," a tool that led to the invention of the "cant dog peavey."

Meanwhile, there are said to be twenty patents under the word *cant hook* in the United States Patent Office, the first in 1856 and the last in 1873. But these were not inventions of principle, only improvements of design and strength.

David Borth (Bayfield) recalls that his father, John, who logged in Waupaca County and north of Antigo in the 1870s, mentioned that the cant dog was still in use when he began logging. This also seems to be confirmed by the article in *Harpers* (1868), mentioned earlier on "The Minnesota Pineries." The writer describes the cant dog as "the old style of a hook attached to a ring on the stock, or the hook with three holes in it, passed through a square mortise in the stock, through which a bolt was passed . . . ."

Eye-bolt cant hook

Ring dog cant hook

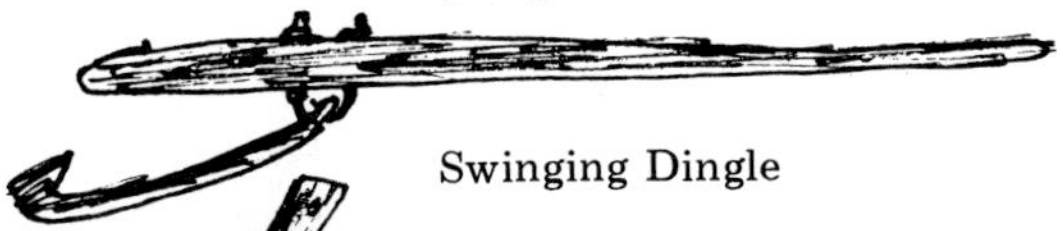

Swinging Dingle

Kanthaken (Ca. 1768)

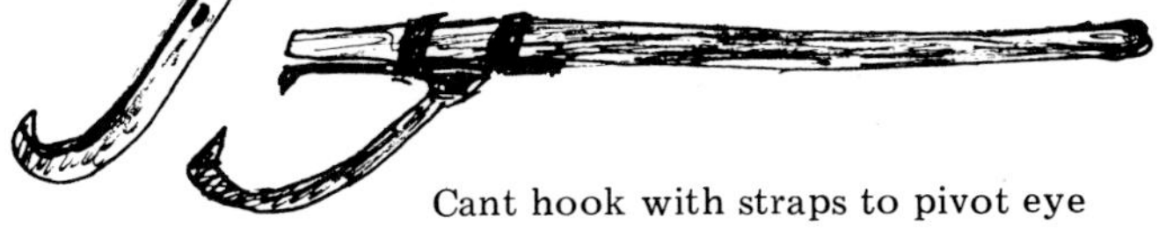

Cant hook with straps to pivot eye

The last half of this sentence is confused, but it seems to refer to a tool known as the "swing-dingle." In one of the photographs from the Whitney Gallery in St. Paul, taken on or near Rum River in Minnesota, in 1864, there is a picture of a crew working on a log drive. Several of the men are definitely using peavies. But in another Whitney picture taken in 1864, two men are unloading logs from a sleigh with what appears to be "swing-dingles". (A copy of this picture was used as an engraving for the *Harpers* article.)

The biggest improvement of the century in the handling of logs was made by Joseph Peavey of Stillwater, Maine, about 1858. This new tool had a hook or dog on it like the ring dog, but the dog swung from a U-bolt or an eye-bolt attached to an iron collar at the base of the wooden stock. In addition, it had a steel spike at the butt which could be used as a jam-pike, or to grip logs with, either in loading operations, canting (rolling) or "tailing down" (rolling logs) on a landing. Oddly enough, for more than a decade after Peavey developed this tool, it was called a "peavey cant-dog," but after it was introduced to Wisconsin, it was called a "peavey" and around the turn of the 20th century, more commonly as "peavey pole." The handle was so much longer than the canthook that the men had the feeling they were wielding a pole as well as a hook.

A blacksmith in Oshkosh, A. Sanford, made the first peaveys in Wisconsin for five or six dollars. There was never any patent applied for on the peavey, and Sanford later began manufacturing the peavies for the mass market and sold them, according to J.L. Whitehouse, a onetime driver on the Wolf River, for $30 a dozen wholesale.

In Michigan, at least as early as 1889, when John W. Fitzmaurice wrote *The Shanty Boy,* the peavey was being called a "pevy (sic) hook."

The peavey was used both in the logging woods and on the rivers from the 1860s to and into the 1880s, and then, some time in the 1900s, the

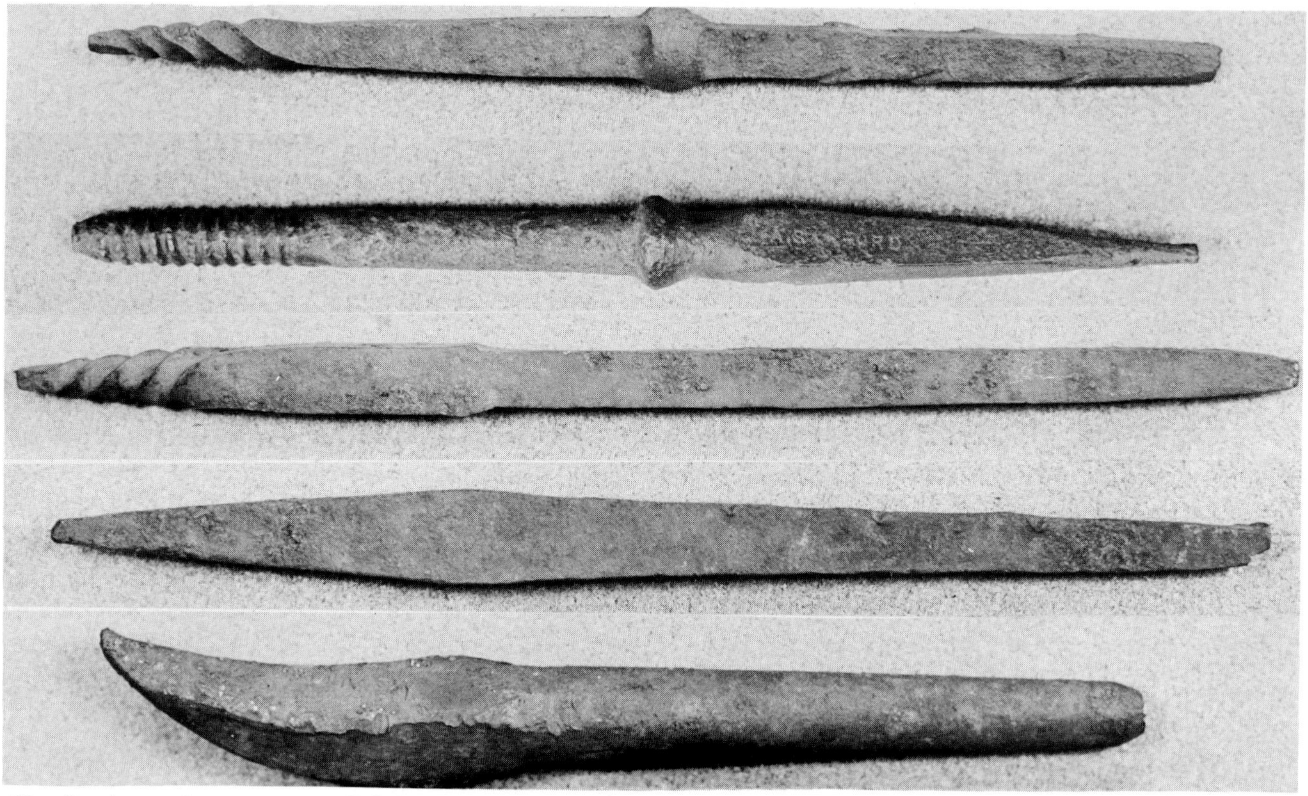

The five irons above are seen slightly less than natural size. The three at the top were used in the stock of a pike pole. The fourth from top was used in the stock of a peavey, and the fifth, or lower one, also in the stock of a peavey but called a "gilray" which some river drivers preferred because it had more strength than the regular peavey point. But the gilray was introduced late in the logging era and not many were made. All the irons above were recovered from former camps in Bayfield County by Howard Peddle of Iron River.

modern cant hook was developed which replaced the peavey in the woods, but not on the drive.

One of the improvements made to the cant hook was an iron snubbing piece which capped the butt of the wooden stock. This had a lip of iron on one side which snubbed into the bark of a log and kept it from slipping. One variation of this was called a "bell-toe," patented in 1887, and made by Sanford in Oshkosh.

The stock on the cant hook was also made shorter than on the peavey which made it easier to handle, both in loading operations and in the sawmill. In loading a sleigh, the man who guided the logs up the skid poles used a cant hook, the top loader used a cant hook, and in the sawmill, the man who rolled or canted the logs up to the headblocks on the saw carriage used a cant hook. He was usually called the "canter," or "canter-on-deck."

All these terms originated in the word "cant" which in the noun form refers to the square, or center part of a log after the slabs have been removed on four sides by the saw. Modern dictionaries fail to include this word as a verb, but, as a verb transitive, it means to "tilt, turn, slant." This can apply to a log which has to be turned and tilted on the headblocks of the saw carriage.

Probably the last addition to the peavey was the "gilray" developed by Robert Gilray, Sr., and patented in 1879. It is described by the late J.C. Lewis, a pioneer merchant at Antigo, who told the Antigo *Daily Journal* (February 8, 1934) that the gilray resembled a cant hook although the stock was longer and heavier. On the lower end of the stock a round, steel spud about eight inches long, was inserted. This had a curved, highly tempered steel point about an inch and a quarter wide which could be used to pry a log better than the peavey pick which was lighter in weight.

The inventor allegedly whittled the pattern for the gilray from a piece of white pine and sent it to Sanford in Oshkosh for manufacture. It was inserted into the end of a peavey like a pick on a peavey only that the swinging hook was hung at right angles to the curved lip of the gilray. This was done to keep the hook from interfering with the gilray when it was being used to pry two logs apart.

The gilray is listed as a "landing hook" by the Sanford Tool Company, but in the woods, or on the river, it was usually called the gilray. Not too many of the gilray type of cant hook were manufactured and they are now a collector's item of considerable value.

A photographer from the Whitney Gallery in St. Paul took this picture on or near Rum River in 1864. The men are unloading at a landing and both appear to be using the swinge dingle type of cant hook. There is no "toe" or socket at the base of the stock which later became a feature of the standard cant hook. This picture and others with sleds on them, all taken in 1864, show that none has a "bunk" across the beam. A picture of a sleigh load of logs, allegedly taken in 1878 in Minnesoata, has a bunk. The evidence of the Whitney pictures suggest that the sleigh bunk was not used before 1864, and probably not until the late 1870s. The Whitney picture above was also used as a model for an engraving in *Harpers*.

The picture was taken at the "horse shanty's camp" of Moore & Galloway Lumber Company, east of Pike Lake (Marathon County), probably in the winter of 1890-91. In the foreground is a logging sleigh, its several parts identified in the overlay. The chain on the tongue was used both to keep the tongue off the frozen ground at night, and at times to lessen the weight on the horses, particularly if they were suffering from sore shoulders.

The crosscut saw

Since time immemorial, trees in the forest were cut down with an ax and once on the ground, the limbs of the trees were trimmed with an ax. The tree, stripped of its limbs, was now called a log and it was sawed into lengths, or "bucked up," which is the more common expression of the American logging woods. In the lake states of America the most common length of a log for general lumber purposes was sixteen feet.

Bucking a log was done with a crosscut saw which was usually five feet or more in length and worked by two men holding a handle at either end of the saw, pushing forward and backward across the log.

One of the of the first times a crosscut saw is mentioned in northern Wisconsin appears in an advertisement of the *Wisconsin Pinery* (Stevens Point) for March 19, 1857. However, these early saws were used only to buck up a log. Finally, in 1876 Henry Disston & sons of Philadelphia, largest manufacturers of saws in the world, came up with a saw that had "cleaning teeth." This was a revolutionary step in the production of lumber because the cleaning teeth, later known as "rakers," made it possible to saw a tree down instead of chopping it down. One old-time sawyer was heard to say that the new saw "cut like a hot knife through butter."

The earlier crosscut saws dropped the sawdust from the kerf on either side of the log because the log was lying on the ground. Now, with a raker, a fraction of an inch shorter that the cutting teeth, it was possible to saw a tree down from the side with horizontal strokes.

According to Allard Peterson, an engineering hobbiest (Elderon, Wisconsin), this new style tooth was placed between the pairs of cutting teeth. It raked out the inverted V of wood and allowed the following pair of cutting teeth to cut the grooves and leave another inverted V which in turn would be

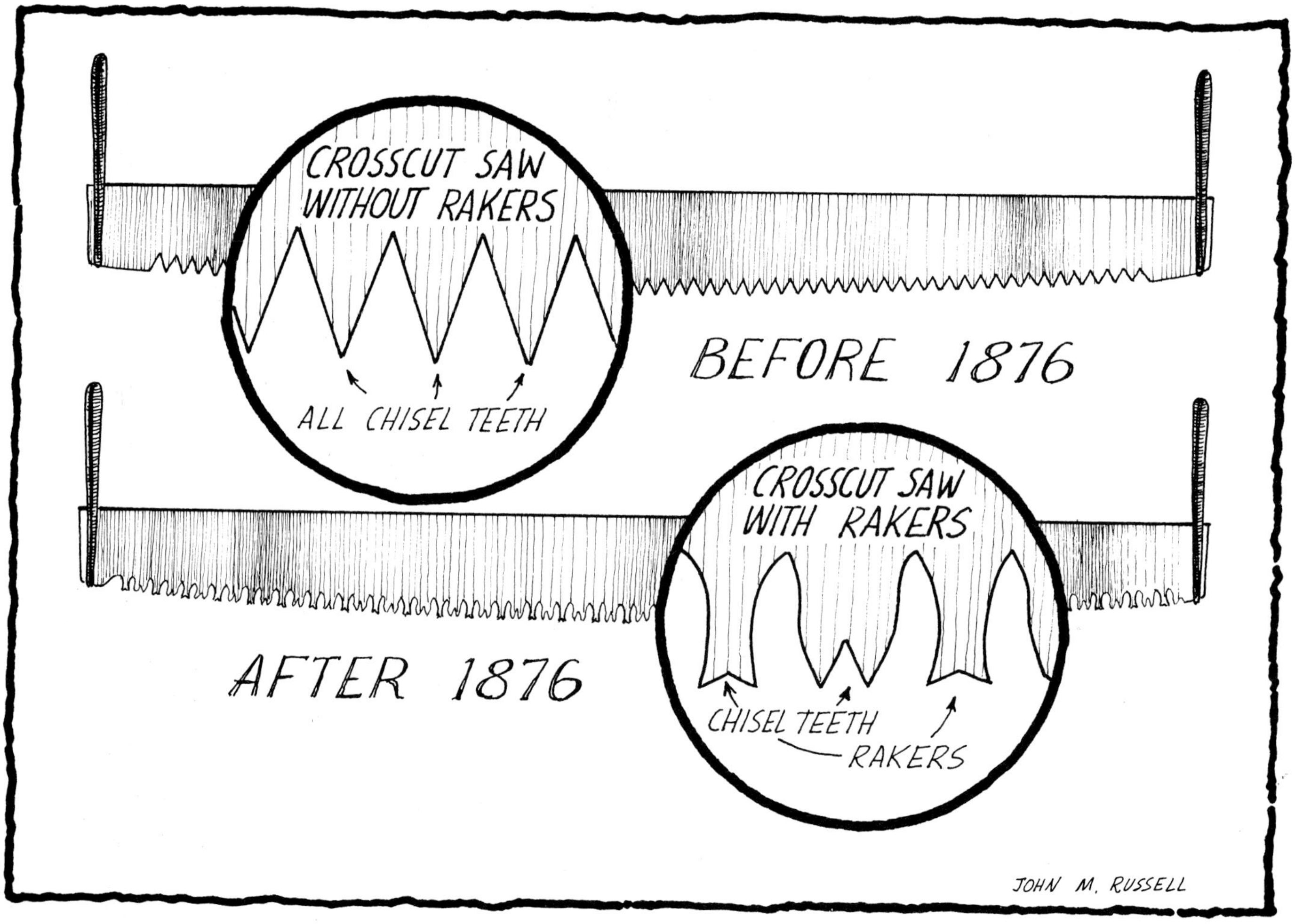

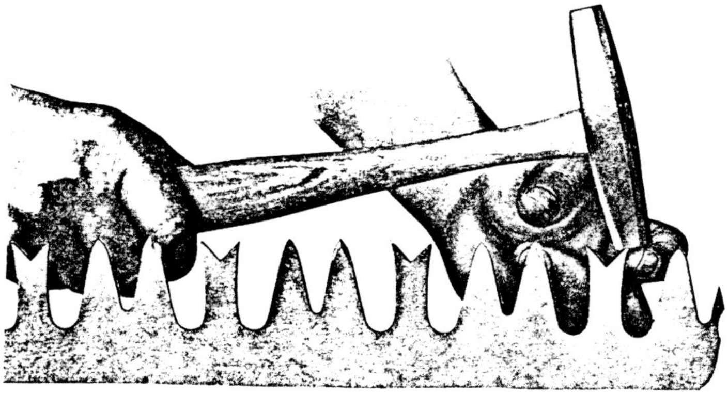

This close-up of a crosscut saw shows the "M" shaped teeth, which are the rakers, and the "W" shaped teeth, which are the cutters. The illustration appears in Henry Disston & Sons catalog for 1914 and demonstrates the proper technique of swaging the rakers.

raked out by the following raker. In this way, he says, the new type of saw made use of three teeth working in a group. One chisel cut a groove on the left, the other cut a groove on the right, while the third, the raker, cut out the center.

Several gauges were made to assist the man who filed the saws, usually called the "saw filer." One, with a small flat file clamped in it, was run the length of the inverted saw. When it had passed over, all cutting teeth were dull, but they were the same length. Another gauge rested on the dulled tips of the cutting teeth on each side of a raker allowing the raker to protrude a bit. It was filed down to gauge level or a few thousands of an inch shorter than the cutters.

The saw filer, who usually had a small shop of his own in camp, used a clamp in which the inverted saw was secured, that is, the teeth of the saw were pointed up instead of down. The cutting teeth were filed at an angle and every tip was brought to a point. All left hand cutters were filed and then the saw turned around and all right hand cutters were filed. With all cutters sharpened, the rakers were last. They stood between the two pairs (see illustration) of cutters that they served with, straight and perhaps a half inch wide. Their end was a shallow V with a flat cutting edge at the outer ends of the V. These were filed sharp. The raker was also filed straight across, that is, flat.

Most saw filers also carried a "rat tail" file with which they made a few strokes of the file between the teeth. This was called "gumming" and, if done a bit each time the saw was filed, it kept the teeth long, and also provided space for the sawdust to ride out of the kerf (or cut) as the saw was pushed and pulled, back and forth.

Once the teeth were all sharpened and the saw gummed, there was still another step, namely, to set the saw, that is, the cutting teeth had to be bent outward a bit so that they cut just a tiny bit wider than the saw thickness. Setting the saw was done with a pressure tool somewhat like a pliers or on a small anvil where the tooth could be bent by a tap of a hammer.

With approximately four raker teeth to the foot, two sawyers pushing and pulling the saw back and forth about sixty times a minute in a thirty inch stroke, about 1200 rakers per minute passed a given point in a log. If each raker took but two thousands of an inch as it passed along, according to Peterson, the saw would cut a kerf two inches per minute. In practice, it was usually faster because the sawyers usually bore down on the saw and the rakers took more than two thousands of an inch cut.

A raker between the cutting teeth of a saw seemed to be a rather simple invention, but why hadn't someone thought of it long before? The Disston company, in business since 1846, today has no idea who invented or introduced the raker into their line of saws. Yet, the raker briefly revolutionized the lumber industry because two men with the improved crosscut could saw down twice as many trees in one day as two men with axes. The "choppers," as these ax men were known, chopped together, one on either side of a tree, one a left-hander and one a right-hander.

But it took more than a decade after the raker was introduced into the crosscut saw before it was accepted by the men in the woods. Choppers were men who prided themselves on their expertise. They could square off a butt log almost as smooth as the new crosscut saw, and they were reluctant to give up a profession in which they had become experts.

However, after 1890, most trees were being sawed down with the improved crosscut, not chopped down. And within the next four decades, most of the good timber in the forest was on the way to a sawmill. In short, the crosscut saw, with raker, had a lifespan of about sixty years before it was replaced with a practical chain saw that one man could handle, about 1950.

Today, the chain saw is leveling the rain forests and teak wood forests around the world faster than they can be replanted. But the chain saw also thins tree plantations and clear-cuts overgrown swamps that would never get cut without it.

Of sleds and sleighs

In Maine, a teamster hauling logs to a landing or to a sawmill used "sleds," and before lumbering became big business, a single sled, called a "bobsled," was also used to skid a log out of the woods, the heavy or butt end elevated onto the sled and chained down, leaving the smaller end of the log to drag in the snow. As time went on, two bobsleds were hooked in tandem to carry the entire log, or load of logs. And this is the way the two sled bobsled came to Wisconsin, where it was also called a bobsled. About 1875, a "bunk" was mounted on top of the beam that held the sleigh runners together, and when this was done, fore and aft, it became a "sleigh."

A photograph held by the Minnesota Historical Society allegedly dated to 1876, shows a bunk on the sleigh. The load of logs is rather large and if that many logs had been decked directly over the beams, the latter would most probably have cracked under the weight.

One of the photographs from the Whitney Gallery taken in 1864 shows what appears to be a Maine sled; it has only beams holding the runners together, no bunks on top of the beams. Thus, it would appear that the bunk was added to the Maine bobsled some time between 1864 and 1875.

In the Lake States, the bunk was bolted to the center of the beam and this acted like a rocker to give the sled a flexibility that the early sleds from Maine did not have, and allowed for a much bigger load both in width and in height. The rear bunk on the logging sleigh in 1901, however, was still rigid, according to Henry Arnold (Green Bay). Not long after he began working in the woods, it was found that by making both front and back bunks into "swinging bunks" that a load of logs could be driven off the road just far enough to let another team pass, although the turn-offs had to be made at specific points where there was room in the woods for a load of logs to pass. In other words, making both bunks into swinging bunks made both bobsleds responsive to the other.

Another improvement to the logging sleigh was made in the Lake States by the addition of the "sway bar," a pole that connected the bunks on either side of the sleigh. These were already being used in 1901 when Arnold started working in the woods. A hole was augered through the ends of the bunks, horizontally, and the bar was run through and held in place by a clamp or pin.

Some teamsters wanted to make the sleigh-bunk nineteen feet wide, but this was considered by members of the state legislature as totally unnecessary. Moreover, a standard had to be adopted, and, while the sixteen-foot bunk was used on sleighs pulled by steam haulers after the turn of the 20th century, most were kept to ten or twelve feet in width.

In Wisconsin, as in Maine and Minnesota, it became a source of pride to be able to excel in the number of logs that could be loaded on one sleigh. In Maine it was called a "crack" load," in Wisconsin, a "star" load. The word *star* came into general use in the first decade of the 20th century to designate someone special, and lumberjacks used it to mean a big load of logs. This was a form of one-upmanship. Every crew tried to out-do the other, and decking a big load of logs—that went nowhere—was one form of this rivalry, but also pride of achievement.

In the early 1890s and down to 1910, a competition arose to see which camp could actually deck the largest load of logs. Perhaps the earliest large load, authentically dated, was one photographed by S.C. Sargent of Taylors Falls, Minnesota, on February 13, 1892, for the Ann River Logging Company in Kanabec County, Minnesota. (The head office and sawmill was in Stillwater.) Four horses pulled 31,480 feet for one mile. The load was twenty-one feet high and twenty feet wide. The runners on the sleds were nine feet long and nine feet apart. The beams were fifteen by eighteen inches square and twelve feet in width, and the bunks (or "rockers" in Minnesota), which rested atop the beams, were sixteen feet and three inches wide.

A year later, fifty perfectly rounded logs were decked on a sleigh at a camp near Ewen, Upper Michigan, which scaled 35,000 feet. This was disassembled and shipped to Chicago where it was redecked and exhibited at the World's Fair. Thus, the Michigan loggers had surpassed the loggers on Minnesota's Ann River. Not to be outdone, loggers in Abe Johnson's Camp (Polk County), not far off

Elias Moses camp (1864) on or near Rum River, Minnesota, shows logging sleigh in foreground with only beams, no bunks or sway bars to support large loads of logs.

the St. Croix River, decked a load that measured 36,600 feet and had a picture taken with two teams of speckled greys standing quietly for the cameraman. But it seems doubtful whether these two teams of horses were more than ornaments because there is no iced road to be observed, and four horses could not have pulled this load. This picture is widely noted in local museums and taverns in northern Wisconsin, and is reproduced in this book.

Finally, to end all competition, the so-called "largest load of logs ever hauled on sleighs by horses" was decked for the Thief River Falls Lumber Company of Minnesota in 1909. It allegedly had 50,589 feet on it and the load was hauled fifteen miles by six horses. But this load was a fraud. For the so-called largest load was not one sleigh, but two sleighs. Moreover stakes, not corner-binds, were used on the sides of the bunks to secure the logs. All this was a breach of ethics among decent, law-abiding top loaders.

However, all these banner loads, beginning in 1892 at Ann River, probably had one thing in common: they must have been decked with a woods jammer or ginpole jammer of some type. A photograph allegedly taken in Minnesota in 1892 shows an A-frame jammer in operation and this dating probably can be accepted in view of the authentically dated Ann River picture taken in February 1892. It is authentically dated because the imprimature of the photographer appears at the bottom of the picture. (This enlargement, twenty by twenty-four and a half inches, in sepia, is held by the Rosholt Pioneer Museum.)

The lumbermen from Maine brought to the Lake States the techniques of using chains—skidding chains, trace chains, cross chains, and loading chains. And the "fid hook" was probably first used in Maine, but most veterans of the logging woods in Wisconsin will not recognize it by this name because they usually called it the "corner-bind." The fid was a special hook which held a separate chain at the four corners of the sleigh bunks. The chain ran through a hole augered at the end of the bunk and when the face, or cornerbind log, was rolled onto the bunk, the chain was wrapped around it and held secure by the fid hook. When the log was about to be unloaded, the fid could be disengaged by striking it with an axe head or even the cant hook stock. The article written by Springer for *Harpers* in 1851 mentions the fid hook.

This was an average load of logs, pulled by two horses, as practiced in the early 1900s. The photograph was taken in 1912 near the camp of Andrew Goli, who was jobbing for Hatten Lumber Company south of Galloway. The logs are hemlock. Elf Goli (left) is the teamster, Tom Boe, top loader, and Andrew holds the rule stick. Fanny is the mare on the off side and Kate on the near side. Goli emigrated to the United States from Norway in 1888 and by 1912 or earlier was his own boss, jobbing for Hatten. C.J. Ruppenthal of Tigerton took the picture but did not get the spelling of Goli's name correct, although no blame need be attached, since many Norwegian "newcomers," as with other ethnic groups, often changed the spelling of their names because they were not certain how to Anglicize them. Nor was anyone sure how to spell "Hatton." This name alternates between Hatton and Hatten, although when W.H. died in 1937, the spelling on his headstone in a New London cemetery appeared as *Hatten*.

John and Edward Henkleman, logging in Marathon County in 1930, used a homemade truck to haul their logs, one of the first logging trucks in the county.

Below, this picture, enlarged from a commercial postcard, turns up in different archives and museums of the state. It may have been taken at Eau Claire, but without any snow on the ground, it seems certain that the lady teamster is not going very far. Nor is she going to manage the horned oxen very well. The photograph was probably posed by an enterprising cameraman.

Hauling cord wood in the snow, probably Douglas County.

This banner or "star" load, as it was called in the early part of the 20th century, was decked near Mellen, and the photograph was circulated for publicity purposes by the Hoyt Drug Store in 1912. Note how the "wrappers" (chains) are bound around the face logs, and how the holding chains run upward and over the entire load. The men at the left obscure the corner-binds. The blank board in the front was ostensibly to give the total footage, but for some reason, was not used. But one team of horses did not pull this load of logs.

The ragged end of a great era in logging in Marathon County is reflected in this picture taken in 1912 at Day Siding, south of Elderon. The two men standing in back row are George Severson and Luther ("Lute") Schofield. Severt Severson is second from the right (in fur collar) and fourth from the right is Thomas Halverson. Alex Hanson, not Mitchell, was foreman, but he is not identified. Inset: John Kluck, who worked as a teamster with this crew arrived a few minutes too late for the picture. To make up for it his picture was taken at his home in Stevens Point in 1979. He said the timber cut that year was all "pick-ups" from blueberry marsh knolls and "scrubby stuff." The crew was loading fourteen cars a day. The main camp of R.D. Hogan, the jobber for Brooks & Ross, lay east of Camp C, or about three kilometers west of Day Siding on the Chicago & North Western line. Henry Dennis was the "walker" for Brooks & Ross that winter.

Picture at left may be the largest load of logs ever decked in Wisconsin. It was taken, in the early 1890s, at Abe Johnson's camp on Bone Lake (Polk County). The road does not seem to be iced and the two teams of horses shown here would not have pulled this load without icing. At the right, a load decked at a camp near Ewen, Upper Michigan, in 1893, was shipped to the World's Fair in Chicago, and redecked. It contained fifty perfectly shaped logs and scaled thirty-five thousand feet. A replica of this load today stands in a public park at Ewen (near Bruce Crossing).

Perhaps the earliest picture of a banner load of logs, authentically dated, was this one photographed February 13, 1892 in Kanabec County, Minnesota, for the Ann River Logging Company of Stillwater. Four horses pulled 31,480 feet of logs one mile. The load was twenty-one feet high, twenty feet wide, and the bunks sixteen feet and three inches wide. S.C. Sargent of Taylors Falls took the picture.

A perfect load of logs, all scaled for the benefit of the cameraman, this Canadian peaker was made up for the J. Arpin Lumber Company (Wood County). Wives and company officers have come out to camp, no doubt on a Sunday, to pose for the cameraman. Frank Minier was head sawyer at the company sawmill in Arpin and may be in the picture.

The Rib Lake picture is enlarged from a postcard in which Brown Studio attempted to show the old and the new as represented by the horse and oxen. The reins are on the horse, however, not the oxen. The farmer is hauling cordwood. On the hillside in the distance stand the stumps and slashing left behind by the lumbermen.

The above photograph was originally nine centimeters in diameter and it was taken with a Kodak Camera. The No. 1 Kodak Camera was manufactured by Eastman Kodak Company with two types of shutters; the model with the cylindrical barrel shutter was announced in June 1888, and discontinued in 1889; the model featuring a sector-type shutter was announced in October 1889 and discontinued in 1897. Both were string-set cameras whereby the photographer pulled a cord on top of the camera to set the shutter. On the back of this picture is an engraving of a long, box-like camera. The picture was taken on January 29, 1892, captioned "Cut Lands," that is, cut-over lands after the lumbermen were done, most probably Knapp, Stout & Company of Menomonie.

On the night of January 15, 1887, according to James Mitchener's novel, *Centennial*, a great blizzard struck the western states. It also left a trail of ice and snow in the Lake States which lumbermen remember as the winter of the Big Crust, or, more commonly, as the winter of the Blue Snow. A heavy snow was followed by sleet that froze on top of the snow, creating a hazzard for both oxen and horses. The picture above was taken at a landing on the Little Wolf River near Norske (Waupaca County) in the winter of the Blue Snow.

When a big crew was in the woods cutting timber, it was common for the saw filer to follow the crew into the woods with his bench and file. This picture was taken near Summit Lake, At left, two saws waiting to be filed or ready for the sawyers.

Two yoke of oxen, harnessed but without bridles, are being used to pull this Canadian peaker. The tugs have been shortened and chains run back to the eveners from the shortened tugs. The teamster stands in front center, holding a goad stick. The top-loader is waiting for one more log to come up before he tops off the load. The cross-haul team, not seen here, would be at the left. The picture was included in a collection given to the author by William Kaiser, son of John Kaiser, founder of the Kaiser Lumber Company of Eau Claire, and was probably taken in Sawyer County.

The icing tankers

Most loads of logs before 1900 were decked in pyramid fashion and were called "Canadian peakers." The pyramid technique no doubt originated in New Brunswick. It was easier to load logs in this fashion, and it did not require as many chains and "wrappers." But it also limited the size of the load. Lumbermen in the Lake States learned to build a perpendicular load of logs, that is, the face of the load on both sides rose straight up, tier on tier, and did not peak at the top. This made possible larger loads, but it also required that the beams on the sled be strengthened by the addition of the "bunk" or "rocker" atop the beam.

But the large load of logs that the men in the Lake States learned to deck, was also made possible by the new technique of icing the roads. Big water tanks, made of planks by the camp blacksmith and carpenter, were used in Michigan in the early 1880s. The tanker rested on the sleigh bunks, and since the sleigh runners were built rigid, specifically for the tanker, the sleigh had two tongues, front and back. In a word, the tanker had no front end. It slid equally well in both directions because the runners were tipped, like skiis, at both ends. When the tanker reached the end of its water supply, the horses were unhitched and brought around to the other end of the tank, which was then returned to a water hole for more water, or taken back to camp. When not in use the tongue was upended and held up by a chain.

To sprinkle the road, the icing crew pulled out two plugs from two holes at each end of the tank, and the water ran out to fill the grooves made by the sleigh runners. The water was also deflected or sprayed nearly a meter outward into the snow by a jet flowing from holes in the center on both sides of the tank, and this created a curb of ice along the length of the road. Icing the roads was done at night to avoid chance meeting with other teams, and also to achieve better freezing conditions when the thermometer was usually below zero.

In the early period, the tank was filled laboriously by hand pails, and then someone thought of using a barrel which could be raised on two skid poles with a team on cross-haul. The barrel slid up the two skid poles and tripped over a hole at the top. One man stood below with a long pole attached to a ring in the bottom of the barrel, and after the barrel was tripped atop the tank, the man below pulled the barrel back and immersed it into the water again.

The tankers were filled at a local creek. In the 1890s dynamite was often used to blast a hole in the lake or swamp. Some tankers had a small stove built inside, fed from a door in the front end, which kept the interior from crusting after it was emptied. But most tankers had no stove. To keep the inside walls from crusting on the return trip for more water, Henry Arnold recalls that he used a gallon jug with kerosene and a wick at the mouth of the jug. He lit the wick, inserted the jug inside the tank, covered the top, and this generated enough heat to keep the inside from crusting.

Above right: At a camp, probably in Lincoln County, this big tanker is being filled with water from a swamp hole. The cross-haul team at the left pulls the barrel, filled with water, up the skid poles where the barrel trips itself and empties into the tank. The man at the right, holding the long pole attached to a ring in the bottom of the barrel, retrieves the barrel and sinks it into the water hole for another "fill-up." The bob sled runners on the tanker are rigid, but curved upward at both ends of the bob sleds.

Water sprinkler for icing roads, Anderson & Bergland Lumber Company, near Iron River, about 1910.

Right: A bitch hook, also called a "goose neck," twenty-eight centimeters long (eleven inches) and in the right hand corner, a double-bitch. These links were used especially to anchor cross chains on the back of a beam on the logging sleigh.

A single bob sled rutter used by the Wall-Spaulding Lumber Company of Big Falls (Waupaca County). A sharp knife was inserted into the runner which shaved the iced ruts smooth for the logging sleighs to follow. But this rutter also has two wooden wheels in back, unlike any noted elsewhere. Jack O'Dell, woods boss, stands at the left holding the handle on one of the wheels. The teamster is Frank Way. The chain below the neck yoke shows that a second team of horses was needed to pull the rutter, but the team has been unhitched in order for the cameraman to get a better close-up picture. C.J. Ruppenthal of Tigerton took the picture in 1914 and used it as a commercial postcard from which this enlargement is made.

Swamp holes and dips in the logging road were often improved by a corduroy (after the French for "cord of the king") road, which consisted of small logs and saplings laid down across a bad spot to make a fairly solid base. Even town roads, well past the turn of the 20th century, were being improved with corduroy but when these roads were later blacktopped, the bumpy corduroy, pushed up by the frost, could still be felt by drivers.

A donkey engine, positioned on the opposite side of the hill, is being used to pull this load of logs to the top. It would have taken three teams of horses to do this, but it was seldom that a logging road went over a hill of any kind. This road was, no doubt, a short cut and a donkey was brought in to pull the loads to the top where horses took over.

By 1918 even this Brisco had invaded the logging woods, this one in Camp 2 for Tigerton Lumber near Summit Lake. George Cook is in the driver's seat; in back, Fred Vollmer (holding dog) and Edward Wollman. A blanket has been thrown over the radiator to keep it from freezing. The back wheels have chains, and on the running board, a tank for carbide lights and a tool box. Two spare tubes are also available. The crank in front was geared in place. A rubber horn is at the driver's left elbow.

One of the first hydraulic lifts ever devised for loading logs was developed by Tigerton Lumber Company and it is shown here in the company yard at Tigerton. The Embarrass River is dammed up to create a pond for booming logs. The logs decked in the background will be sacked into the water and sent to the mill, not visible but at the right.

The "Lima Shay" engine

The last major development that hurried the depletion of the Wisconsin Pinery was the invention of the "Shay" engine. Before 1900 railroads were building spurs into the woods off the main lines to bring out the timber for shipment to mills. But there was a limit to the number of cars that a narrow gauge engine could pull up a grade. Ephraim Shay of Haring, Michigan, noted the problem. The rods that ran from the pistons to the driving wheels on the standard gauge locomotive did not provide steady traction, and the wheels began to spin on a grade of more than one percent, that is, one foot of rise to one hundred feet of track. In 1880 Shay developed a steam engine in which a geared shaft ran from the engine to each wheel on the track and meshed with the ring of gears on the wheels. No matter how hard the engine might pull, the wheels, geared in this manner, could not spin.

The final patent was issued on June 14, 1881, and in the next sixty-five years the company works at Lima, Ohio, manufactured 2,770 "Lima Shay" engines which were sold to lumber companies throughout the nation. It was used in switching yards as well as in the woods on the narrow and standard gauge lines.

A small "Shay" was used between Ingersoll and Eland Junction in the 1890s, affectionately named the "Henry Sherry," after one of the partners of the Ingersoll Land & Lumber Company. According to Allard M. Peterson, (Elderon), a railroad buff, these Shays carried all their weight on their drive wheels, including the weight of the tender. When the engine back-geared to the drive wheels through a system of corner gears and universal joints, the exhaust from the engine was so great that for anyone accustomed to listening to a standard gauge locomotive, the approach of the "Shay" sounded like a runaway engine.

An offshoot of the Shay principle of a gear shaft from the engine to the wheel was picked up by a man in Maine, Alvin Lombard, who expanded on the principle of the cleated tracks first used by the Holt Tractor Company, about 1900, to power a steam log hauler, the ancestor of the modern snowmobile. The driver of the steam hauler, who was called the "headlight" or "steerman," sat up front on a bobsled that could be turned by a wheel resembling a car wheel. The engineer in the cab back of the boiler kept the machine running and the fireman kept it hot. Altogether, sixty-seven Lombards were made. A restored model, the

While the McGiffert loader works in the background, the Lima Shay or "geared engine" waits for the last flat car to be loaded before high-balling for the sawmill in Rib Lake.

only one in the United States, apparently, is on exhibit at the Lumberman's Museum in Patten, Maine.

In 1903 the Phoenix Manufacturing Company of Eau Claire, Wisconsin, took over the Lombard patent and began to manufacture its own version of the steam hauler called the "Phoenix." More than a hundred were made in the next decade, and, while most of them remained in Wisconsin, a few found their way to camps in Minnesota, Michigan and the Rocky Mountains. It found no buyers in the Pacific Northwest where there is not enough snow for good sleighing.

The sleighs behind the steam hauler for hauling logs were usually wider than the average, with twelve to sixteen-foot bunks, and the engine could draw up to fifteen sleighs in tandem, with four to five thousand feet on a load, according to icing conditions, level of land, etc. One who remembers a steam hauler near Elderon (Marathon County) is Albert Elliott, who, as a youth, heard the shrill whistle of the engine in the far distance while it made its way from "Camp F" (west of Elderon), six miles south to the "slashings siding" (northeast of Galloway) on the Chicago & North Western Railway. Here the logs were loaded with a McGiffert and shipped to Fond du Lac.

Mr. Elliott recalls that the heavy "ping-pong" sound of the Shay gears, mingled with the crunching noise of sleigh runners on icy roads in the dead of night, seemed to shake the entire forest around. But this particular steam hauler was used only two or three winters. In the spring of 1910 it was driven between two skidways of logs on the "slashings siding" and parked there for the season. In July a great fire swept through Franzen township, destroying millions of feet of timber on the stump as well as skidways decked at various landings including the one at Galloway. And with it went the steam hauler. It was never replaced. Out of the one hundred or more manufactured in Eau Claire, only two on exhibit in Wisconsin are fairly well preserved, one at the Rhinelander Logging Museum, and the other on the grounds of the Public Museum at Wabeno.

Withal, the steam hauler, was an invention of limited capability. The railroad spur or the narrow gauge superseded it nine times out of ten, and by 1915 these spurs and narrow gauges had penetrated almost every township in some northern counties. In the final stages of the great pine harvest, "boarding cars," built on flat cars, were used to provide housing and mess facilities for the crews of the Vilas County Lumber Company. These cars could be moved around, a camp could not, and in this instance the company found it cheaper to build a few boarding cars and spot them on a railroad spur than to build camps.

The Thunder Lake Lumber Company, located on the north side of Rhinelander, may have operated the last logging railroad in the state. One of the branches of the Thunder Lake line ran from Rhinelander east, across present Highway 45, and one branch ran north somewhat along present Highway 17 to camps near Sugar Bush. A narrow gauge, it was terminated in 1941.

One of the conveyances for moving lumber was the "big wheel skidder," developed in Manistee, Michigan, by Sylas C. Onespark in 1870. This had wheels so large and an axle so high that it skimmed over the tops of stumps and underbrush. It was not meant for the long haul, but for bringing a cluster of logs, wrapped in one chain, out to a logging road. But this too was an "invention" of limited capability, both because it was rough on the horses, and because little summer logging was done in Wisconsin.

Not many photographs of the "big wheel skidder" in action have been found. One was taken south of Shell Lake in the 1890s and is shown in an accompanying reproduction. The Rhinelander Logging Museum has a pair on exhibit; Camp 5 at Laona has a pair, and a third is on exhibit in the public park on Highway 2 in Ashland.

A steam hauler for the Fountain-Campbell Lumber Company at Donald, once situated east of Cornell (Chippewa County).

Gagen, once a community lying between Rhinelander and Monico Junction, was the main camp and sawmill of the Gagen Lumber & Cedar Company. The steam hauler at the left is pulling at least fifteen loads of logs scaling, apparently, 80,000 feet.

Close-up of the wheel and seat for the stearman on a steam hauler, this one is preserved on the grounds of the Public Museum at Wabeno.

Another view of the steam hauler seen in the preceding page for Fountain-Campbell Lumber Company. The stearman leaned over the wheel with both hands to turn the traction gear and chain hooked to the bob sled in front.

Steam hauler of the New Dells Lumber Company of Eau Claire, probably logging in Sawyer County, about 1910.

A steam hauler, probably Chippewa Valley, pulling string of empty logging sleighs. This picture affords a close-up of the holes that were augered through a bunk. The one on the outer end was for the sway bar. The next one in, where the chain runs through, was for the corner-bind. The fid hook can be seen lying atop the bunk in center. Dangling at each end of the beam are two bitch hooks or "goose necks." In the distance at the right stands a water tanker for icing roads.

Picture of steam hauler reproduced from a postcard which is postmarked at Couderay, November 6, 1917. The card is addressed to a lady in Westby, and the note says: "I am well and working but I gess (sic) I will quit, so don't write this time." J.M.

Steam hauler for G.M. Jones Lumber Company, leaving Camp 3, located east of Wabeno (Forest County) winter of 1907-08. Up front on the stearman's seat is Edward Moore. The engineer, James Ellswood, stands in front of the cab, and William Janke, fireman, in door of cab. Standing below, left to right, are Naurenus Nelson, Jack Bath, Tom Lambert, Emil Thaves and Mike Moore. Sitting or standing on the first load of logs are Edward Soderstrom, John Gayhart (holding snowshoe) and Christ Knutson. The Jones sawmill was located on the North Branch of the Oconto River at Wabeno. The steam hauler seen here stands on the grounds of the Wabeno Museum, the only complete steam hauler left in Wisconsin.

The Holt Tractor Company used the "cat" principle to develop a tractor that replaced the steam hauler in the woods. This, however, did not completely eliminate the need for horses. This picture was taken for Tigerton Lumber Company at Summit Lake, early 1920s.

In the Green Bay logging district, lake schooners were used from Marinette-Menominee, Oconto, and Green Bay to ship lumber to cities on Lake Michigan, especially to the big lumber yards in Chicago. Here is the *Lucia A. Simpson*, last of the three-masted schooners. Built in 1875, she carried lumber and forest products for fifty years. Charles Goodman, whose family founded the Goodman Lumber Company, donated this picture to the Marinette County Museum.

Railroad logging

What the coming of a railroad meant to the logging industry is no better illustrated than by the picture below of eight men, all leading lumbermen or investors, who are waiting on the dock at the mouth of the Menominee River at Marinette for the arrival of William B. Ogden, mayor of Chicago, and a director of the Chicago and North Western Railroad. These men wanted the railroad extended from Green Bay to Marinette-Menominee because shipping lumber by lake schooners to Chicago was slow and expensive.

All across the northern counties of Wisconsin, lumbermen waited for the railroad. In the beginning, they were satisfied to be on a railroad line. For example, with the arrival of the Wisconsin Central in Stevens Point in 1871, there was no need to raft lumber to markets on the Mississippi, a hazardous, costly and slow procedure. It could be shipped out by train in cars without number. Moreover, before the railroad came to Stevens Point, the sawmills were sawing mostly undressed, that is, rough lumber. It did not pay to raft dressed lumber down the Wisconsin River.

With the coming of the railroad, planing mills could be built to dress and finish the rough lumber in the same yard. There was little waste and no shrinkage. And a trainload of lumber could be shipped out every day.

But the railroads changed the means of transporting not only lumber, but the logs deep in the forest. Spurs, both standard gauge and narrow gauge, were built off the main lines to bring out the timber, not in big loads, but small loads, faster and more often. As soon as one timber belt was exhausted, the trackage was pulled up and laid down in another lumber belt.

One man whittles, the others wait patiently for the arrival of William B. Ogden from Chicago at Menekounee in 1867. Left to right: John Spalding (the youth), Judge E.S. Ingalls, Jesse Spalding, Nelson Ludington, Daniel Wells Jr., Doctor J.C. Hall, Abner Kirby, Isaac Stephenson, and Chaney Simonds.

Above: A landing on Wood Lake (Taylor County) for Rib Lake Lumber Company. Robert Hess stands in cab gangway. Straddling logs, second from left, is Bert Aiken, foreman at Camp 4, and next right, Chester Curran. Photo looks south.

Left: A logging train for the Vilas County Lumber Company approaches a grade on a logging railroad probably en route to Winegar in Presque Isle township.

Below: "Passing Camp 6 at Fosterville," says the caption inscribed on the picture by A.J. Kingsbury, photographer from Antigo. The people on the flat car at the right are either visitors or lumberjacks going to town for the day. The attitude of the men at the left is relaxed. This is no doubt Sunday. Fosterville, named after J.J. Foster, once lay in Presque Isle township (Vilas County) and Camp 6 was part of the operations of the Vilas County Lumber Company, which failed in 1910. It was taken over by William S. Winegar of Grand Rapids, MI and the name of the community was changed to "Winegar." A hotel and boarding house and eighty company houses, all painted either white or green, were built for employees. The company had three hundred and fifty men in the local mill and in three camps within a few kilometers of the mill. The Chicago & North Western railroad came down from Mercer to pick up the lumber sawed and dressed at the mill complex in Winegar.

George Morris, engineer, stands on footboard of this logging and switch engine at Ashland.

There was once a village called Nashville situated a few kilometers east of Pelican Lake. A railroad spur ran from the Chicago & North Western line at Pelican Lake to Nashville to pick up lumber and bring in supplies for the mill crews and camps of The W.H. Rogers Lumber Company.

A load of logs on cars for the Robbins Lumber Company of Rhinelander, about 1905. Car stakes on flat cars were still a novelty, but this train of logs has been secured first with chains. On the front of the engine is an early version of a small snowplow. At the left is George Smith, chief engineer for Robbins & Jack Mylrea. The picture was taken at Columbus Lake (Oneida County) where Mylrea was woods boss.

Engines, buried in snow, fight their way through the drifts to open the tracks for the Thunder Lake Lumber Company near Jennie Webber Lake (Oneida County), about 1925.

Belching smoke, this Northern Pacific logging train arrives at Brule (Douglas County). Log train came from "Brown's Camp" on Brule River. This picture dated 1898, is one of the first to show that stakes, even though short, were being used to speed loading operations on flat cars.

The long logs seen here were used for piling, or for telephone poles, etc. They were being loaded on a spur running between Whitcomb (now a ghost town) and Morris (Shawano County) on the Chicago & North Western Railroad. The young man in felt hat at left sits on a velocipede which followed the railway tracks.

Above: Brooks & Ross of Schofield were logging northeast of Bowler before World War I, and the landing shown here is presently the site of the Stockbridge-Munsee Community Headquarters. (Below): a long train load of railway ties leaves Tripoli (Lincoln County). The ties were sawed in the local mill of Yawkey-Bissel Lumber Company. Their main mill, however, was at Hazelhurst. Picture was taken January 29, 1924. (Upper right, opposite page), a Chicago & North Western train approaches Koepenick Siding, north of Antigo, mail pouch suspended on the arm waiting to be scooped up by one of the crew as the train passes through. The slatted platform construction between the fences was called a "cattle guard," a device to keep the track open but to discourage cattle because their hooves would get caught in the slats. The depot at Koepenick Siding was an old boxcar without wheels. (Center right, opposite), Boulder Junction (Vilas County), in 1910, looking west. Brooks & Ross cut the pine in this part of Vilas County, and their office is at the right. Dennis Paquette operated a bar and hotel in the large building down the street, and across the railway tracks (left) stands the first depot of the Milwaukee Road which was built specifically to accommodate Brooks & Ross who were bringing their logs in from camps to the north and east of Boulder and shipping them to the company mill in Schofield. (Lower right, opposite, logs behind a boom on the Wisconsin River at Tomahawk with passenger car and caboose of the Marinette-Tomahawk & Western railroad.

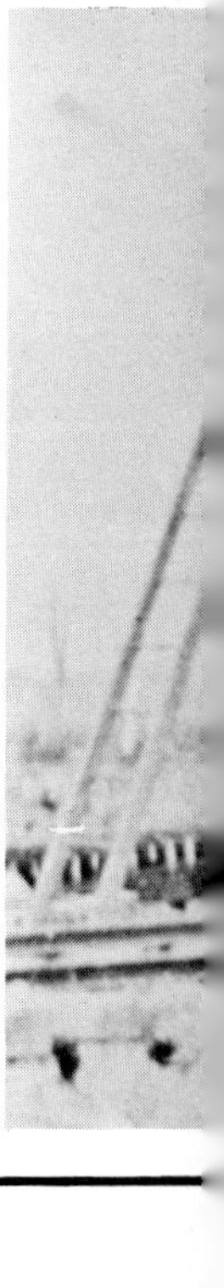

KOEPEN
SIDING

MARINETTE TOMAHAWK & WESTERN RY.

Here is a crew in the woods peeling and bucking up hemlock. The man at the right holds a "spud," a tool with an iron, shovel-like snout with which the bark is pried loose and peeled off in strips four feet long. The next two men are using a cross-cut to buck up the log, and the owner, probably standing in the background, acts as straw boss.

Loading hemlock bark on a siding near Wausau. In the 18th and 19th centuries the bark peeled from hemlock trees was worth more than its wood. It was sent to tanneries where the juice was squeezed out to make an acid used in tanning leather. The man standing on the dray is probably a farmer. His heavy fur coat hangs over the end gate. This type of dray, commonly used for hauling cord wood, hemlock bark or fire wood, rested directly over the beam on the bob sleds.

These men are using axes and broadaxes to make eight-by-ten inch timbers from hemlock and tamarack. The log is "scored," that is, notched by the man in the rear, Henry Kersten. The others are using axes and broadaxes to smooth out the intervals between the scorings. Albert Seefeldt stands up front (left), broadaxe poised in air. The picture was taken in Wyoming township (Waupaca County), about 1915.

Loading railroad ties into a box car at Wausau, about 1910. The men are using "lug-hooks," also called "timber carriers," to lift the ties into the car. The man at the left wields a cant hook.

A flume, filled with running water, once carried logs through the hills west of Drummond to the big sawmill of the Rust-Owen Lumber Company in Drummond. This appears to be the only place in Wisconsin to use this technique, although it was common in the mountains of the Pacific Northwest. The company discontinued the flume before 1900 and this picture was taken many years after it was abandoned and cattle are grazing in the cut-over land. The edifice in the distance was once a railroad trestle.

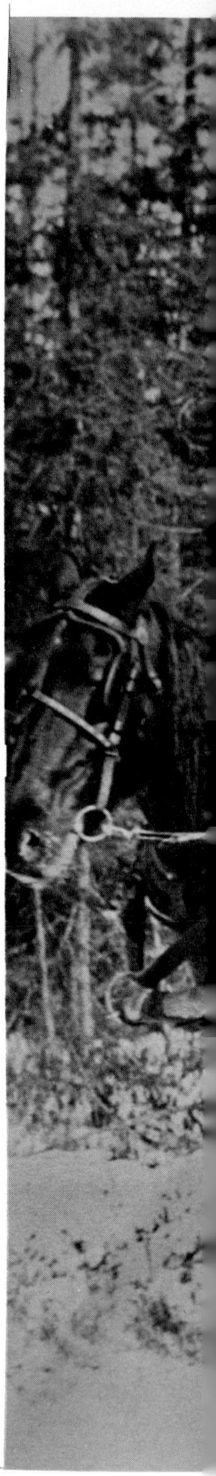

A load of eight-foot pulp logs is being hauled to market on a very cold day near Big Falls (Waupaca County), about 1915. The corner-bind dangles in front of the bunk but is not in use. The stakes are secured by chains running under the logs from front to back.

A crew pauses on a foot bridge on the Little Eau :aire River, town of Easton (Marathon County) en ute to cut and peel hemlock for Louis Mortenson ho was jobbing for the Wright Lumber Company of errill, probably summer of 1908. The man at the ;ht, holding crosscut, is James Carbine; third from ;ht, Phillip Kumba; fourth, Edward Carlson, fifth, ed Hutton (holding cant hook), and at the far end left, Henry Larson, who was cooking for the crew. ated in front is John Mortenson.

Steam hauler for Anderson & Bergland Lumber Company at Alvin, Wisconsin (now a ghost town) but located near Iron River. Logs, which appear to be mostly white pine, are being hauled to railroad siding for shipping out, probably to Hines sawmill in Iron River. Man at left behind horses is probably "tote" teamster hauling supplies to camp from railroad head. He has baled hay in rear of sleigh box, barrel and other supplies up front. Whippletree lies on stump in foreground and sacks behind stump can only be guessed at. Photo taken about 1910.

Ox teamster holds goat stick over oxen with one hand, cant-hook over shoulder in other. Oxen are skidding logs out of forest to nearby skidway. Yoke over necks of oxen can be seen behind the horns. Chain to log was secured to hook under center of yoke. After 1900 horses replaced oxen in the wood because a larger breed of horses had replaced the drivers and pacers of an earlier period. Style of clothing worn by crew suggests photo was taken before 1900. Picture was taken on West Fork of Black River for Island Mill Lumbering Company.

Above: Decking logs on the skidway with A-frame jammer (upper left). Here the logs are being skidded, not rolled up the skid poles, by a crotch chain attached to a cable on a traveling block running back to the jammer. The logs appear to be mixed pine and birch and several have dry rot.

Below: the first railway depot at Tomahawk. Two scantlings atop the ridge board apparently held the telegraph wires, but the condition of the shingling suggests that this picture was taken some years after the building had been abandoned.

Life in camp

Logging crews moved into the woods to work about the 15th of November and stayed until the first of April. A few men went home for Christmas, but many had no home to go to. Sunday was a holiday for everyone, and it was used to clean up and to sharpen axes.

Most camps in Wisconsin operated in the winter to take advantage of the snow and ice to haul bigger loads of logs than could be hauled on wagons. But there were companies that logged the year around, too. In an advertisement appearing in the American *Lumberman* on April 21, 1900, Brooks & Ross of Schofield said their mill was running day and night and their crews were working in the woods the year around.

Life in camp in the period from 1840 to 1870 in Wisconsin was primitive. Log shanties, or "bunk camps" as they were called in Michigan, were built with little attention to comfort or warmth. They were built low to the ground, the outer walls only three or four logs high, most logs probably a meter in diameter. The floors were usually puncheon. The door to the shanty, the only door, was located at one end facing either a river or spring, and it was usually so low that a man six feet tall had to stoop to get in, at the same time that he stepped over a threshold, hewed from a stringer wide enough to match the door jamb. In this instance care had to be taken by the carpenter to match the threshold at the bottom of the door, or all manner of small creatures would be looking for a place to come in from the cold.

There were no windows, as a rule. The light was a skylight over the fireplace in the center of the room. The lid could be lowered when the weather was bad. And since the crews were small in this early period, the men slept and ate in the same room, the kitchen area usually in one end and the bunks in the other.

For lighting at night there were kerosene lamps, and lanterns. It sounds gloomy today, but the men who worked in the woods had nothing better at home either. Moreover, they were not reading in bed at night. Most of them were too tired after working from dawn to dusk to do anything but eat and "roll in."

It is commonly accepted as a fact that the log shanties built by the loggers were comfortable and warm. Actually they were warm around the stove, or fireplace, but the interior walls, when the temperature dropped below zero, were usually covered with frost in the morning. Although the logs in the wall were large, not enough attention was paid to fitting them, one atop the other, to

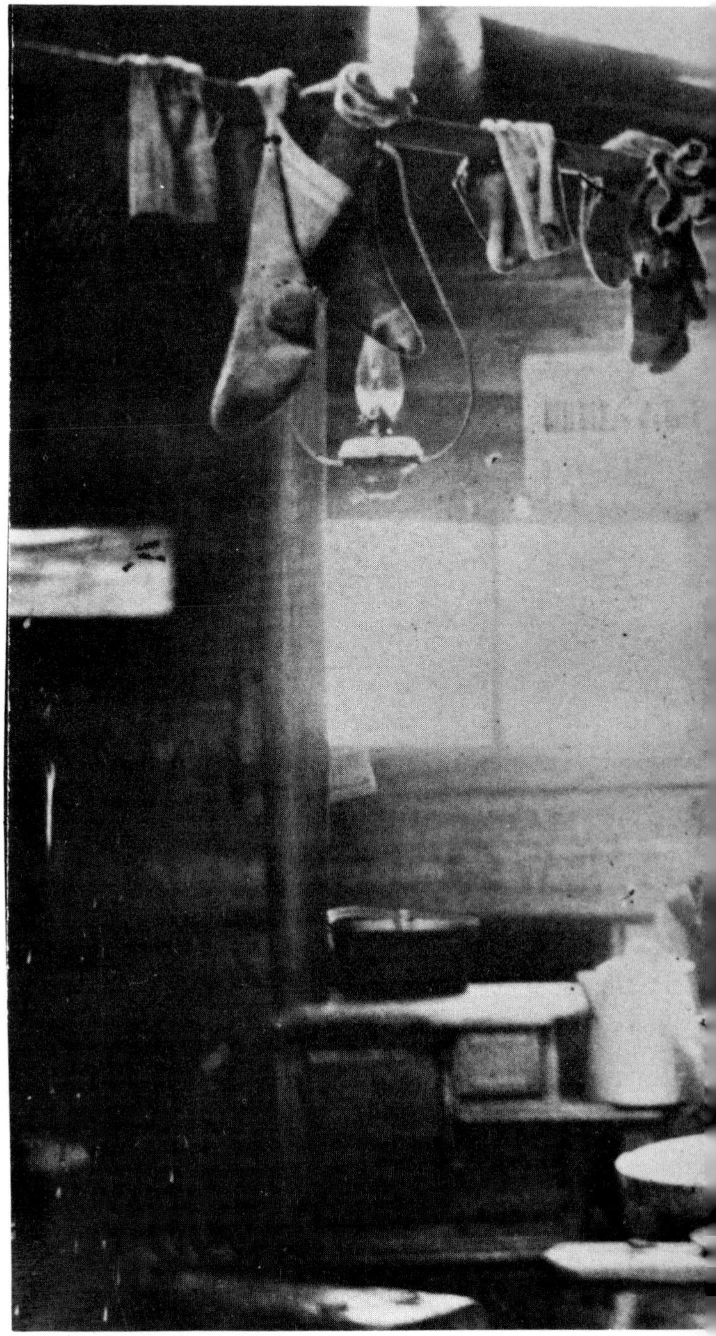

keep the air out. Moss and mud were used for chinking but this froze and cracked and seldom remained sealed throughout the winter.

And contrary to popular believe, the log houses that the pioneers built were not cozy little homes, warmed by a picturesque fireplace. Most of them had no fireplace. They were cold and miserable except around the stove, and there was nothing anyone could do about the floor; that was always cold.

Louie Blanchard, in a biographical sketch, *Lumberjack Frontier*, describes the inside of a sleeping shanty in these words:

> Along the sides there was a lot of racks to hang sock and clothes on to dry at night. Everybody's got wet from the snow every day and we needed a lot of racks for drying. When the socks began to send off steam at night, you sure knowed you was in a logging camp. When you mix the smell of wet socks with the smell of baked beans and chewing tobacco, you have a smell that a lumberjack never forgot even if he lived as long as Methuselah.

But Blanchard fails to distinguish between early and later woods camps. In 1886, as a picture taken near Antigo reveals, the camp was divided into separate eating and sleeping quarters joined by a "dingle" or storage area between the two buildings. The entrance to both buildings faced each other across the dingle. This was one of the most common architectural arrangements in the woods for the next thirty years in Wisconsin and Minnesota. The horse barn, the blacksmith shop, and the store or "wanigan," were usually in separate buildings, whether of log or frame construction. (The Menominee Logging Camp Museum at Keshena features two log buildings connected by a dingle.)

Log buildings were usually roofed with shake shingles covered with hemlock or cedar boughs to keep the heat in, or the roofing consisted of split basswood logs, hollowed out, which were laid one up and one down, called pantile roofing, similar to Chinese and Spanish rooftops of porcelain tile.

The earliest log shanties had a fireplace, but not at one end of the room with a chimney. Instead, it stood in the center of the room where it

This high-ceilinged log hut served as kitchen and sleeping quarters for a crew in the camp of Elias Moses on Tibbets Creek, a feeder stream of Rum River in northern Minnesota, in 1864. The kitchen range, pipe rising straight to the ceiling, stands in center with kettle boiling and coffee can on the underpart of the stove. The two men in foreground are seated at hand-made benches in front of hand-made table. In the ceiling hang socks and clothing, and in upper left, a kerosene lamp. The sleeping bunks are probably at the left. The light for this rare picture came from a window behind the cameraman who was probably from the Whitney Gallery in St. Paul.

threw off the most heat, and gave the most light. It was embedded in a large sandbox eight to ten feet square in two to three feet of sand. In the more elaborate fireplaces, posts, about eight feet high, with overhead stringers, were built at the four corners of the sandbox. A fire was maintained in the center and a big iron kettle hung over the fire on a swivel attached to one of the corner posts. The cook kept a long shovel to juggle the logs around the fire, to remove ashes, or to dig a hole in the ashes for a kettle or earthen jar to simmer beans in overnight. The French-Canadians called this a *cambuse*, which appears in some English texts as "camboose," but a word found only in the French dictionary. The smoke from the fire escaped through a light in the roof directly over the fireplace. But the *cambuse*, like the ordinary fireplace, wasted fuel, and gave off little heat in proportion to the amount of wood it burned. And the bunks the men slept in were too warm near the fire and too cold away from it.

In the United States, kitchen ranges and wood heaters replaced these crude heating arrangements in the early 1870s, but in the Canadian "bush" (woods) there were camps using the *cambuse* until past the turn of the 20th century.

Sleeping bunks inside the shanty in the early period were built of hewed logs, or planks, and the sleepers slept with their heads to the wall, feet to the fire. These bunks were called "breechloaders," presumably because they could only be entered or crawled into from the one end. A breechloader rifle is loaded from one end—like all guns.

After heaters replaced the fireplace, the inner walls of the sleeping shanties did not frost up as badly as they once did. Also, bunks were built parallel to the walls, and, as a counterpart to the breechloader, were called "shotguns." But the common term was "bunk" and in most camps two men shared each bunk.

Throughout most of the logging era, there were no springs, only boards to lie on, covered with hemlock boughs or straw, or both, one blanket over and one under, and a pillow made of anything that could be rolled up, like a wool shirt or coat. By 1924 at Eagle River mattresses were laid over the board bunks instead of hemlock, and the men slept one to a bunk, not two. In the 1930s, even spring beds were being used in some camps.

In a letter written from Little Bull Falls (Mosinee) on April 17, 1855, to the Wisconsin *Pinery* (Stevens Point), a correspondent (referred to elsewhere in the book) remarks about his good health. He attributes this to the "good quarters, warm blankets and plenty to eat." He said he was living in a log house which had "three apartments, a tremendous large chimney with two fireplaces, one in each room capable of burning from one to twelve feet [sic] wood . . . at all events a glowing fire in a family fireplace." After having done a hard day's work, he said "we collect around the brisk fire, telling our anecdotes, singing our songs, and likely discussing some principle in philosophy; we are a cheerful and happy set . . . ."

It seems doubtful that this correspondent was typical of the "shanty-boys," most of whom had no time for philosophy. Moreover, the writer of this blurb was living in a camp on the Wisconsin River, not far from supplies and provision in Stevens Point. In the logging woods, camps were isolated and difficult to get to. "Tote roads," to bring in supplies, were little more than trails

Oxen were seldom used on cross-haul for decking on the skidways, but here a yoke of oxen is straining mightly against the long decking chain stretching from the top of the skidway down to the evener. These are large oxen and explains why they could be used.

through the forest big enough for a yoke of oxen to squeeze through.

Albert Stuntz, a timber cruiser on the Namekagon River in 1868, did not find much joy or entertainment in the camps he visited. According to his diary, he spent a day at a camp of the Mackey Brothers, not far from the future city of Hayward, and complains about the "verry [sic] long day [with] nothing to read or amuse myself about camp. Life is carried on as usual, some mending boots, and fiddling, some reading, some sleeping & some talking."

The crowded, dingy conditions of a logging camp of the Civil War period is suggested by one of the Whitney Gallery photographs taken near Rum River, Minnesota, in 1864. However, there was no need for more elaborate arrangements. The crews that went into the woods before the 1870s were small, often members of the same family, five to ten men working for a jobber or a sawmill owner who had limited facilities and small assets. There was not a big market for lumber products before the Civil War, and larger crews would have outrun demand.

Early photographs show that barns, or "hovels" as they were called in the 1850s, were of log construction, scarcely high enough for a horse to stand in, although the oxen managed. The door to these hovels was usually so low that the horses had to learn to duck their heads to get inside. In the 1880s and 1890s, when horses began to replace oxen, the hovel became a "stable," often a frame structure, or at least partly frame, at which time it became a "barn." In the larger camps with many teams of horses, one man oversaw the cleaning of the barn and general care of the horses. He was "barn boss" and his word was law inside the barn in the same way that the cook was lord of the kitchen.

One chore the barn boss did not have to worry about was currying the horses. Each teamster curried his own horses, fed and watered them. This was almost a sacred rite because the teamster took pride in the appearance of his horses, argued about them, and lied about how smart they were.

The Ingersoll Land & Lumber Company (Marathon County) built what was probably the largest barn in Wisconsin. It stood on the north side of present Highway 29 in Section 2 of Elderon township. Three eyewitnesses, interviewed independently, vowed that this was the "largest barn in the United States . . . maybe the world!" It was allegedly more than one hundred feet wide and three hundred feet long, larger than a modern football field. It had three floors, the first for the cattle (oxen), the second for horses, and the third for haymow. The second floor, laid with heavy planking, inclined toward the center for drainage.

The barn was commonly referred to as "Ingersoll's Folly," after Ira S. Ingersoll who began logging in Marathon County in 1880. When the mill at the village of "Ingersoll" burned in 1897, Ingersoll sold his timber and moved away. The barn was demolished and parts of it probably went into a new barn built for an Indian mission conducted by the Norwegian Lutheran Church Synod on the one-time Ingersoll farm.

Most logging camps after 1875 had their own blacksmith shops. After 1900 some camps hired a carpenter too to help keep things in repair. A good blacksmith, however, could make just about anything in iron or in wood, and when he was not

After the lumberjacks left their old camps and moved farther north to new logging areas, the buildings they had abandoned were usually occupied by homesteaders, anxious to start farming in the cutover land. Here a lady, surrounded by stumps, has planted a garden.

busy shoeing horses, or repairing sleighs, he was making cant hooks and peaveys.

Finally, every camp had a foreman, or "woods boss" who ran the camp, saw to the bookkeeping if necessary, kept track of the board feet of timber cut every day, hired and fired, and paid off the men. (A board foot measured one foot by one foot by one inch.) In a big camp, he was usually assisted by a clerk or bookkeeper. In the larger camps the foreman also had his own quarters and, if married, he usually brought his wife to camp. If he had no family, the scaier often shared the foreman's quarters and perhaps acted in his spare time as a clerk. There were no rules about these things. Everything worked according to circumstance and need. In some camps the office of the bookkeeper served as the camp store or "wanigan."

When there was more than one camp crew cutting for the same lumber company or jobber, the foreman was superseded in authority by a superintendent, usually called the "walking boss," or "walker," because he often walked from one camp to the other to keep an eye on things.

This type of camp construction was widespread in Wisconsin as late as the 1880s and, in isolated cases, into the 1890s. It combines stable (left) and sleeping-eating shanty (right) in one building, with dingle in the middle for storage. The picture was probably taken in Douglas County, and the time is either late fall or early spring. The man at the left, pipe in mouth, balances a crosscut saw over his shoulder, and his partner, an axe.

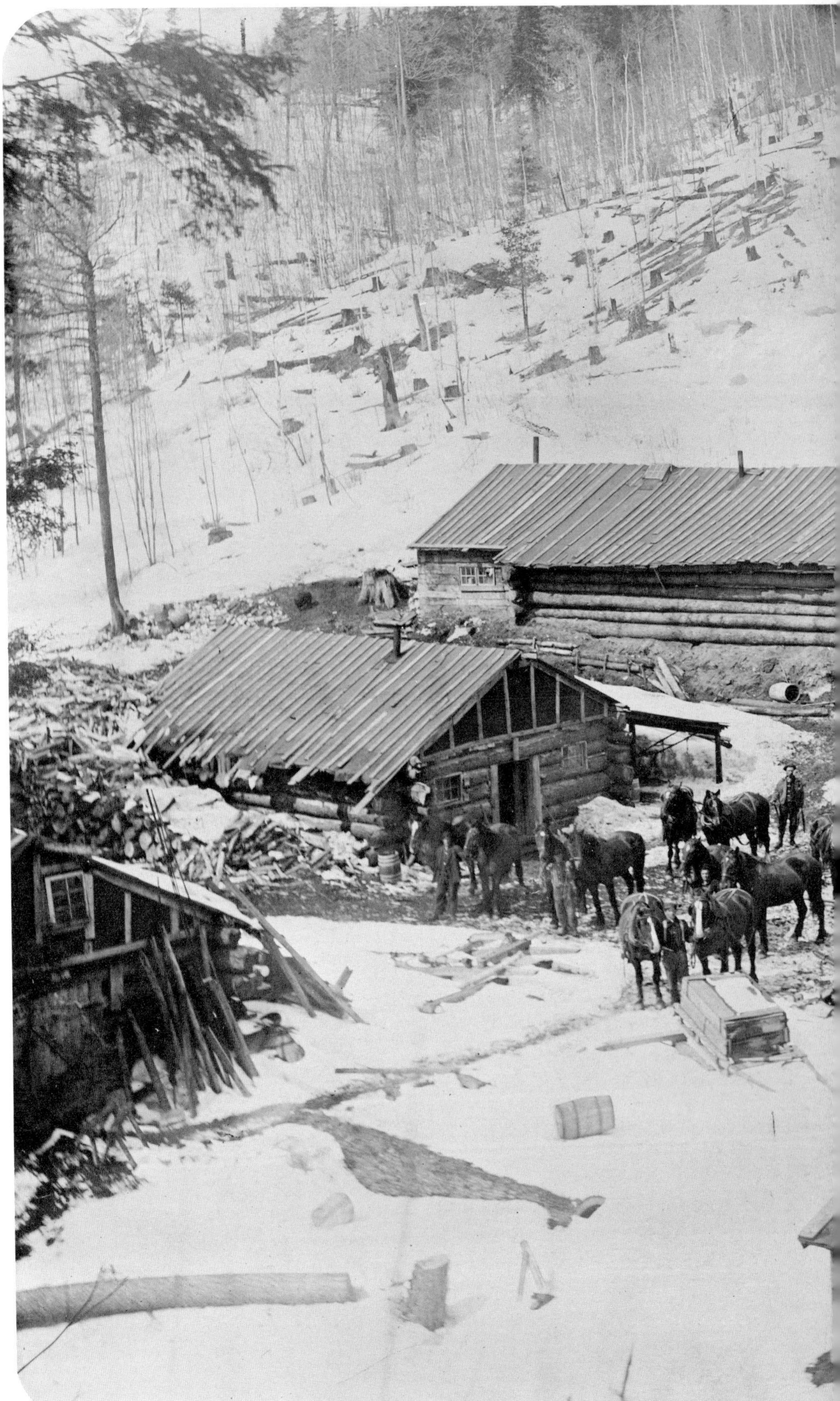

Camp buildings such as these replaced the earlier and smaller camps in the 1890s, but still retained some of their features. The long building on the hill no doubt represents sleeping shanty on one end and kitchen area on the other, with dingle in the middle. (A building similar to this has been built at the logging camp museums in Eau Claire, Keshena, Rhinelander and Wabeno.) The small building at lower right may be the saw filer's shack, and the building at left center may be the blacksmith shop.

The choreboy, or "lunch boy," seems pleased by the attention he is getting from the cameraman and wants his mule to share in his pleasure. The sled, often called a "jumper," was used to haul containers of hot food into the woods for the crews at lunch time. The log building in the back was, no doubt, the kitchen, and it shows the use of tar paper for siding and roofing which began in the logging woods in the early 1900s. Bundy Lumber Company was cutting around Parrish (Langlade County) at the time.

This was probably the largest woods camp ever built in Wisconsin. It stood on Pigeon Lake, west of the present village of Drummond (Bayfield County), and it was built in the 1880s by the Lammers brothers, Albert and George, who logged off most of the pine around Pigeon Lake. According to Gordon Sorenson, the company sleigh-hauled their logs to Upper Eau Claire Lake, and, by raising the level of a small river running out of the Upper and Lower Eau Claire lakes, managed to float their logs into the St. Croix River, and from there down to Stillwater where the Lammers had their mill. The picture was taken in 1893, probably on a Sunday after a new snow.

A woods crew, probably for Kaiser Lumber Company at Camp I, east of Winter (Sawyer County), some time between 1905 and 1910. One of the cooks holds a horn, called the "gabreel," which was used at times for calling the crew to dinner. The youth, with hand on back of dog, is the choreboy, and the head cook at the left of the choreboy is James Rice. The building at left seems to be the stable and is covered with balsam boughs for roofing. Building at right, with light in roof, is probably the kitchen shanty.

LIBBEY & READ'S Camp
Feb, 1886
PHOTOGRAPHED BY— MORGAN BRO'S
ANTIGO, WISCONSIN.

The late T.J. ("Tommy") Thompson, drove a big dog into the logging camps of the north, selling watches and jewelry. Below, he arrives in camp. These two postcards were presented to the author by Thompson at Barron in 1961, not long after he retired as a director of the First National Bank.

Only three photographs by the Morgan Brothers, William and Charles, are currently available, two held by the Langlade County Historical Society, of which this is one, and one by Mrs. Milton Page of Seattle. The stable in this picture is probably at the left, kitchen shanties at right. Some of the men hold crosscut saws and cant hooks. In a letter dated February 16, 1963 to Earle Holman from Charles L. Reed (not Read) of Fond du Lac, Reed states that the firm of Libbey & Reed was his uncle, Daniel Lord Libbey, and his father, Rufus C. Reed. They were logging for Williamson & Libbey Lumber Company who had a sawmill on Fox River in Oshkosh. The logs cut near Antigo were sleigh-hauled probably to the West Branch River and from there drivers pushed them down the Wolf to Bay Boom, and, after sorting, across Poygan to Lake Winnebago. Libbey & Reed dissolved partnership prior to 1895, and when Charles was a boy his father was cutting timber around Mercer. The Morgan pictures appear to be the oldest authentically dated pictures of logging in northern Wisconsin.

Camp of R. Connor Company, probably **north of** Stratford, around the turn of the 20th century. In front, one foot **extended, stands** Mike Baltus, woods boss for **many years** with the Connors.

Everyone has been treated to lunch, probably, and the visitors are ready to leave. In the cutter are Mr. and Mrs. Ole Emerson of Chippewa Falls who are probably visiting one of Emerson's camps east of Cable (Bayfield County). The other couple are Mr. and Mrs. Olaf Sherman and son. Sherman was a watch and jewelry salesman.

No talking at table

While there was no sign on the wall forbidding it, every lumberjack knew that one of the unwritten laws of the woods was "no talkin' " at the table. With thirty to fifty men in a mess hall, everyone eating elbow to elbow as at a church lutefisk supper, the cooks would not have been able to communicate with one another if talking had been allowed. They kept an eye on the serving dishes and coffee cans and filled them as soon as they were empty.

The serving dishes were set up to feed four men, that is, two men on opposite sides of a long table, seated on benches, so there was no need to pass anything. The "boarding-house reach" was acceptable, and most men speared a piece of bread or just about anything on the table with their fork. But no one stirred his coffee with his thumb. Perhaps it happened once, a long time ago, somewhere, and it became the inspiration for a folk ballad about the "Frozen Logger" who, in the first verse of the ballad, went into a small cafe one evening and—

A 40-year old waitress
to me these words did say:
"I see that you are a logger
and not just a common bum,
For nobody but a logger stirs
his coffee with his thumb."

Meanwhile, an eye-witness account of a cook at work in an early logging camp in Wisconsin is provided by Mrs. J.L. Washburn whose father, Joseph Pattee, took a contract to log off a tract of timber north of Stevens Point in the winter of 1870-71. There were eight men in the crew besides Pattee. One of the men insisted on bringing his wife along, and as a result she was hired as a cook. But she was pregnant and soon found it difficult to keep up with her duties. Alma Jane Pattee, not quite fifteen years old, had to come from school in Stevens Point to help her father. She described the logging camp as a "very rude building" which was divided into a large and small room. The large room had a big fireplace, bunks in the corners, one above the other, a long dining table with benches on either side, and two or three chairs. The other room had a cook stove, a bed, a sofa and a pantry where the food and dishes were kept.

Alma Jane had never made bread before, but she found by experimenting that she could make a fairly decent substitute out of "sourings" and soda. By sourings, she meant something that "pancake batter would be after it had stood for a while and

Camp 9 kitchen for Bundy Lumber Company near Parrish. The older man at right is probably head cook, and other two his cookees, one holding the "gabreel." One poster on wall advertises "Frank Marharas Big Minstrels," and at right, a drama called "A Pair of Country Kids." A paper bell hangs in the ceiling, probably a decoration left over from Christmas dinner. The kitchen area may be seen through the door, and despite the cracks in the floor, it is swept clean.

turned sour." She had no eggs or milk, and she could not send for any. She followed a recipe found on the baking powder can, omitted the eggs, and used water instead of milk which was "every bit as good as milk when baking powder is being used." (Quoted from her book, *Letters to My Children*, privately published, ca. 1945.)

Salt pork in this camp and every other camp in Wisconsin was the basic meat diet in 1870. It was basic because the pork, salted in brine, kept in warm weather as well as cold. Besides serving pork, Alma Jane Pattee baked beans in a big earthen jar which she buried in the fireplace and banked with coals before she went to bed. This was a tried and tested method of preparing beans in camp, and most accounts agree that there was nothing that tasted better than beans simmered overnight in an earthen jar, and presumably flavored with "blackstrap" molasses. In Maine, they were called "logging berries," a nickname that did not catch on in Wisconsin.

Although Alma Jane tells about making bread for the crew, most cooks at the time made biscuits, not bread. Bread required butter, corn syrup or lard as a spread, whereas biscuits could be eaten without any. The basic diet up to 1870, then, was biscuits, salt pork, baked beans, molasses, and tea. Even potatoes were scarce.

A fairly good idea of what the "shanty-lads" were eating in the spring of 1857 is provided in a book account kept by Burton Millard at the store of James E. Armstrong in Wausau. On March 11, 1857, Burton made these purchases, almost surely for a crew working either in the woods or at a local sawmill in Wausau.

39 lbs. pork	$6.63
barrel of flour	5.50
barrel of flour	6.00
49 lbs. dried apples @ 16¢	7.84
24 lbs. tea @ .06	18.10 (?)

This picture of a camp kitchen was probably taken at Star Lake and the cook in the center is believed to be Adolph Swenson from Wausau. In some camps, the kitchen area was isolated from the mess hall, as noted on the preceding page, but here kitchen and dining are in the same room. The dish washer's table and tubs are at the left, towels drying over the stove where towels have dried since time immemorial.

37 lbs. coffee @ 16¢ .... 6.66 (?)
25 lbs. coffee @ 20¢ 5.00
12 lbs. saleratus @ 11¢........... 1.32
37 gal. syrup @ 11¢......... 50.84 (?)

The bookkeeper, as seen in the above, was a bit confused in his figures. However, on March 20, Millard was back in the store for thirty-four pounds of sugar @ 14¢. On May 21, he purchased one bushel of potatoes for $1.00; on June 24, he purchased one half gallon of molasses for 70¢; on July 2, one pair of shoes for $2.50; and on July 6, 110 pounds of dried apples @ 20¢.

This account shows that both tea and coffee were being purchased, although most accounts of life in camp agree that very little coffee was served to the crews before 1880. It is also clear why molasses was not as popular with Burton as it came to be later: it was too expensive. There is no mention of prunes in the account.

(Burton Millard, who came to Wausau in 1849, went off to war in 1862, and was killed in action.)

Up to 1880, there was little variety in the menus whether for breakfast, lunch or supper. "Hot cakes," a favorite breakfast dish later on, were not being served, yet, a store in Kilbourn (Wisconsin Dells) was advertising in the Wood County *Republican* on March 23, 1871, that it was ready to supply the "river trade" with potatoes, pork, beef, flour, beans, ham, eggs, syrup, tea and tobacco, all of which suggests that the river raftsmen ate a lot better than the men in the woods.

At a *grocerie*—a combined food and liquor store—at Wisconsin Rapids, the owner, also in March, 1871, was offering 1500 pounds of "good roll butter . . . very cheap . . . for lumbermen."

Roll or rolled butter was butter wrapped in some grade of parchment or wax paper, but if it was "very cheap," it was probably very bad, too. However, this advertisement shows that butter was available, although not many cooks used it in camp, not only because it was in short supply, but because it was hard to keep. Everything that came into camp had to be measured against the one yard stick: would it keep? But the river raftsmen were not worried about this. They purchased foodstuffs as they sailed down the river, stopping overnight sometimes in towns where fresh eggs were on sale.

In some logging camps the foreman built a "root cellar" to preserve fresh meat, milk, vegetables. It was built into a sidehill, covered over with a gabled roof to keep the rain out, and boarded in front to keep the animals out.

Most camps in the 1880s and 1890s had their own cow to supply fresh milk, not for drinking, but for baking and cooking. Even as late as 1924, Moore & Galloway at Galloway was not serving anything but skim milk for coffee, and margarine as a bread spread.

Tea was the standard drink at all meals up to the 1880s, although, as noted in the Millard account above, some crews were also getting coffee. After 1880, most camps served both coffee and tea, and conditions in camp and the food as a whole improved. By 1900 it was a rule of thumb in the woods that no food was too good for the jacks.

Some idea of the kind of food the men ate in camp in the late 1870s and into the 1880s is provided by the late Mrs. John Moran (nee Ada Cottrell) who, with her husband, cooked for H.C. Scott, a jobber on the Wolf River. She told a reporter of the Antigo *Journal* that for breakfast she served fresh pork, beefsteak or sausage, griddle cakes, fried potatoes, doughnuts and coffee. Lunch, better known as dinner, included beef stew, or pot roast, potatoes, onions or rutabagas, brown gravy, bread and butter, pie or pudding, and coffee. She said the butter was usually "horrible," meaning, no doubt, rancid.

Supper included fried potatoes, cold meat or fish, batter cake, prunes, mince or prune pie, and coffee or tea. Most bunkhouse historians agree that prunes, which were bought dried, were the favored fruit dish whether cooked for sauce or made into pie, or as one raconteur recalls, "ground up for meat balls."

There is reason to believe, however, that Mrs. Moran's memory of the food she served in the 1870s was influenced by what she was serving in the 1890s. Fresh pork or beefsteak, for example, were rare in the 1870s because there was little refrigeration except in winter. She also forgot to mention that dinner, or lunch, was usually served in the woods where the men were working, not in camp. The food was brought out to the crews in a compartmentalized container strapped to the back of the lunch carrier, or hauled out in a single horse sled. The lunch carrier built a fire to warm up the tea or coffee, but the food, which was supposed to be hot, often froze, not because it was not hot, but because the tin plates the food was served on were ice cold. To keep ahead of the cold, the men ate fast, standing up, or seated on a windfall. There was no hour off for lunch, but twenty minutes at the most with scarcely time for a smoke.

In some camps, according to Dale Peterson (Eau Claire) in a Ph.D. dissertation on logging in the Chippewa Valley, venison was served. One hunter in two seasons on the Eau Claire River in the late 1860s allegedly shot eighty-five deer and two bears. Poaching out of season was not unheard of, and some camp foreman encouraged the men to spend their spare time hunting. But this was rare, mainly because most of the men who came to the woods in the 19th century had no rifles, Daniel Boone to the contrary.

The lumberjacks who worked in the woods after the turn of the 20th century, lick their lips when they think of the good "johnny cake" they were served for breakfast, or supper, usually with molasses. John Goul, writing in *Ford Times* (August 1979) believes that johnny cake originated with the French on the St. Croix River (between Maine and New Brunswick) who got yellow cornmeal from the Micmac Indians living along the south shore of the gulf of St. Lawrence , and made *Jaune-gateau*, literally, yellow cake. "Later," he continues, "when johnny cake . . . became a nourishing hotbread in the lumber camps, the Yankees called it yellow jack—jack being Downeast for a batter. Gingerbread was blackjack, griddle cakes were flapjacks."

By the 1890s the camp cook had become the unchallenged boss of kitchen and sleeping shanty and he was wearing a white apron. His word could only be countermanded by the foreman, and not many of the latter dared to risk the wrath of a good cook. A poor cook lasted only a few days before he was told to "draw his time" and given his "walking papers."

If the foreman would not fire a bad cook, the crew found ways to encourage him. At the Mohr & Stotzer sawmill at Holt (Marathon County), a cook had been brought up from Portage, home office of

Three fiddlers, three cooks and probably the foreman pose for this interior shot taken, about 1902, at the August Mason camp near Brill (west of Red Cedar Lake Barron County). The walls are plastered with old show bills, mostly minstrels, the one of the man at the right advertising 'A.W. Martin's Show," and at the far right the "Georgia Minstrels," and "Uncle Toms Cabin." Pinup art from the "Rialta" vaudeville circuit and others add to the color.

Loggers eating at the Gilkey and Anson camp, probably near Three Lakes. Dan Delaney, back to camera, was foreman. This picture was no doubt taken with flash powder. The men ate in semi-darkness because there was not much light given off by the lamps. No one did any talking except the cooks, a tradition started in Maine and carried over to the Lake States and to Canada. The abundance of good food made up for lack of light and conversation.

the lumber company. But the men did not like his cooking. The late Joe Mienke told the author that "we jumped him because we didn't get what we wanted. Overfed us on liver, bought in barrels. Some of the men poured kerosene in the barrel."

The late Joe Stencil, a lunch carrier and bull-cook at Holt about the same time as Mienke was working there, told the author the following:

> I never forget that time. Walking through that big swamp [carrying the lunch container]. Oh! It was hard. No snow that winter to speak of, and the swamp so soft. And I was bull-cookin' for Joe Cherwonka and he'd sleep late and I had to be up first thing in the morning and build the fires and even make the pancakes for the first table. I peeled potatoes, carried water, carried wood and everything. For breakfast it was pancakes, pork and sometimes beefsteak, but mostly pork, and sometimes wieners. For supper I remember we often had prunes and if they didn't eat it all, we squeezed the pits out and made pie the next day. For butter we had oleo, but I couldn't eat the stuff. I put drippings on my bread.

Mrs. Moran recalled that it was the custom of the cook to have plenty of pie, cake and doughnuts on hand for Sundays to treat the men as well as visitors, such as friends from another camp, or an itinerant preacher, or salesman passing through with a satchel full of cheap watches and jewelry. In most camps there were also hoboes passing through. It was risky to deny them a meal. Arson was one of the most feared criminal acts in the woods and a tramp scorned might set fire to a barn filled with horses, or even to a forest in order to "get even" with the "bosses."

Mrs. Moran also failed to mention how the crews were called to supper in the evening. A long horn, resembling a Roman instrument of ancient times, called a "gabreel", (after the angel Gabriel) was used in some camps, and whenever the crew had a picture taken in front of the shanty, one of the cooks came out and posed with his "gabreel."

There seems to have been no pattern for the use of the horn. All of the photographs currently available, which include a horn on them, were taken in the 1890s and early 1900s in the northern counties of the state, not in the central Pinery. The use of the horn depended on local conditions, such as the relation of camp to where the men were working. It was not used as a wake-up call. The men were awakened either by the cook or bull-cook with a loud shout, or, if a woman was cooking, by an ear-splitting scream, "roll out!" or "come an' get it!" The expression "daylight in the swamp" was heard in Wisconsin, but not too often. In the 1920s, and 1930s, as the end of the logging era approached in the northern counties, a "gong" made either of a tire rim or an angle iron, was struck by the cook to wake up the men or to call them to meals.

At least an hour before the sawyers, choppers, or swampers were wakened, the chore-boy, or "bull-cook," nudged the teamsters. It was important for them to feed, water and harness their horses because as soon as the sawyers were finished eating, the skidding teams had to be ready to accompany them into the woods.

Sunday was wash day and also for boiling clothes to reduce the vermin, mainly "grey legs" or lice. This picture by Kingsbury was also taken at Camp 9 for Bundy Lumber Company. The Bundy sawmill was located on Pine Lake in Lincoln County, a community founded in 1891 by D.K. Jeffris who dammed up Pine Creek to make a pond and built a mill on the east shore. The Jeffris mill burned February 9, 1895 and was rebuilt, but two years later, burned again, at which time Jeffris sold his mill property and timber to the Bundy Lumber Company. The mill was rebuilt and continued until 1912, when the Bundy mill was acquired by the Larson Lumber Company. The latter ran the mill another decade. Highway 17 skirts the east shore of Pine Lake where the mill once stood.

Sunday in camp was devoted not only to cleaning up and washing clothes, but also boiling clothes to reduce the vermin population. No one ever thought of taking a bath because there was no place to take a bath and if there had been, it would have been too far to carry water for fifty men. In the early 1900s many of the jacks in the woods were heads of families who usually went home on holidays, but ordinarily, a person can go a long time without a bath in cold weather without feeling any ill effects. In fact, many of our pioneers, who had no experience in the logging woods, were just as opposed to bathing as the lumberjacks. They thought it was unhealthy to be bathing too often.

Usually, there was a wash basin at one end of the sleeping shanty, with a barrel of rain water or creek water to dip from. In fall and spring the wash bench stood outside. Soap and towels were furnished, but family men often brought their own towels. No one ever brought a toothbrush, no one wore glasses, and only a few could afford a dollar watch.

Sunday was also a day when an itinerant lay preacher might arrive in camp—just in time to eat—and call a meeting in the sleeping shanty. The men listened out of courtesy, if nothing else, and when the hat was passed, they donated to the "good cause," at which time the visiting preacher went on to the next camp, knowing full well he hadn't accomplished very much.

There was no drinking in camp, not for moral reasons, but because some men never knew when to quit, and when drunk, became violent or abusive. The men lived in close quarters, and violence of any kind could upset the peace of the bunk house. Gambling was also frowned on, especially poker, and a game with anything higher than penny ante was discouraged.

The precautions taken to keep order in camp

Sunday in camp for Dells Lumber & Shingle Company of Eau Claire. The picture was taken at Camp I, east of Winter (Sawyer County) in January 1908. At the left is a grindstone which is being used to sharpen an axe. The container below the stone holds water to keep the stone wet. Above hangs clothing, mostly socks, for drying. The man at right leans back for a shave, probably by the chore-boy acting as barber.

and discourage heavy gambling followed two decades, roughly 1860 to 1880, when there was much greater freedom, for one reason, because the crews were smaller. Ola Johnson, who lived in Wisconsin Rapids, gave an interview to Thor Helgeson of Iola, some time before 1914, which appears in Helgeson's book, written in Norwegian, called *Fra Indianernes Lande (From the Indian Land)*. From the nature of his comments, it seems clear that Johnson's recollections of life in camp went back to the period between 1860 and 1880. A translation of the Johnson interview has been made by the author, and it follows in part:

There were many nationalities in camp. Yankees, Germans, Frenchmen, Irishmen and we three Norwegians. In the evenings we had a big time. The Germans and the French and Irish competed with each other in singing their national songs. The Germans had many kinds of songs and popular numbers, one of them the old hunting song called *Im Wald Und Auf der Heide* (In Forest and Meadow). The Frenchmen sang the Marseillaise and some war songs; the Irish sang songs like 'An Irishman Lives as a King in his Land with Whisky and Pipe.' The Yankees also sang a few songs, usually national songs. But we Norwegians were poor singers and we often sat there like three owls among a flock of crows.

Cards were played nearly every night and all day

Sunday. There was neither holiday nor holy day. Swearing in German and in English we heard day and night. The first words we newcomers from Norway learned, as a matter of fact, was how to swear in English . . . .

While the lumberjacks spent most of their leisure time in singing, card playing and dancing, that was not all. There was also violin playing and dancing, since there were no less than four fiddlers in the crew. But real musicians were lacking. In a logging camp we were not particular about the quality of music.

There was a 'grub hall' where the log jobbers stored their rafting grubs [grubstakes] for rafting lumber, and this building was usually empty the first part of the winter. Here the boys and girls, men and women, gathered to dance the waltz and French-four far into the night. Many came from town on Saturday night. They often brought whisky which they treated or secretly sold. Disturbances were frequent in the lumber camp and in the 'grub hall.' People drank, danced, swore and fought so that on the following morning, the hall looked like the worst kind of pig sty. Some people acted worse than heathens, the only difference being that the lumberjacks did not eat those who got knocked down.

The next village from our camp was a lumber town, and in a town like this there were, no doubt, some decent people, but many of them were plain trash. At least that's the way it was in this town sixty years ago. But at that time it already had a great number of men and women of shady character, and among them, shame to say, were two Norwegian girls.

When we were through with our strenuous work in the woods, we were fortunate enough to get our pay, for pay was uncertain in those days. Many men worked a whole year without getting a single cent. We returned home and found our wives in good health although now and then they had found it difficult to get food and wood. The wood we cut in the fall had not lasted longer than the Day of Our Lady *(Mariamesse)*, but you can be sure Nils had big eyes when his wife said, 'There's your New Year's present.' It was a pretty little girl!"

There was practically nothing that the men would not do to embarrass a newcomer or "greenhorn" in camp, anything for a laugh. Nils Theodor Larson, of Iola, remembers an incident which also appears in Helgeson's book:

The choppers always had to shout in a loud voice whenever a tree was ready to fall in order to warn the sawyers and others nearby to clear out. One time we had a star chopper from the Wolf River who wanted to be different. He ground his axe every night and smeared it with fat. On this particular day the chopper was working on a tree in the same neighborhood where I and Ola Holla were sawing on another tree. The man did not give us any warning but we heard a cracking and swishing and down came the tree! Ola jumped to the side but I had to throw myself along the log we were working on. Stumps and sticks and branches rained down, and I was knocked out. I did not regain consciousness until Ola Holla dragged me out of the snow. Even then, as serious as it was, there stood that fool and laughed. But then the other men became angered and he was forced to pack up and leave.

Another story, on the lighter side, also from Helgeson, goes this way: John Aas, a Norwegian settler in the town of Alban (Portage County) was doing some logging on his farm and had hired a Norwegian newcomer boy to help with the swamping. One day the newcomer saw a porcupine in a tree and asked his employer (speaking in Norwegian) "What kind of animal is that?"

"That's a bear cub," replied John.

"How much can it be worth?"

"About five to six dollars."

"I will cut down the tree and take it."

"Then you'll have to be quick about it."

The newcomer cut down the tree, grabbed the porcupine across the neck and got his mittens covered with quills.

"Aa hutte metu!" (good grief) he exclaimed, "The bears in America are sharp!"

"That's why they are so expensive," replied John.

The ethnic background of the men who worked in the logging woods of northern Wisconsin included French-Canadians, "Blue Noses" from Nova Scotia, Irish, Germans, Scandinavians, and Yankees, the latter mostly from Maine and New York state. But the "Irishers," as they were often called, dominated the French-Canadians and all others. William Hoyer, of German descent, grew up among the Irish at Chippewa Falls and was along with them on the *Chippeway* drive. His manner of speech could easily be mistaken for an "Irisher." Polish lumberjacks were numerous in Marathon, Portage and Oconto counties, and after 1900, Finns played a prominent role in camps along the northern tiers of counties butting on the Upper Peninsula of Michigan.

Thermometers abolished

An early account of lumbering which appears in *Harpers* mentions that the foreman of a crew, preparing to go into the woods, stops to look at the thermometer which, presumably hung outside the door. But around 1870, it seems, all thermometers, large and small, were abolished in the logging woods. It was one thing for a small, family-size crew to be concerned about the weather, but in a big crew, knowledge of temperature readings thirty below zero might intimidate some of the less robust. What they didn't know wouldn't hurt them. But the forest sheltered the men working among the trees. It was on the landings and skidways during loading or decking operations that the wind could whip around the corners and cause the whiskers of a man's face to frost up, and the nostrils and bellies of the horses to show white. Still, no matter what the chill factor—an expression unheard of—there is no record of any crew that ever stalled because of the cold. Deep snow, or no snow could hold up operations, but not the cold.

The men were tough in body and in mind, and they were hungry. Many of them were immigrants, ready to suffer almost any discomfort to get a job. Oral tradition tells of some Polish immigrants in camp near Bevent (Marathon County) who, on arrival, were so poor that they could not afford wool socks inside their boots. But they defied the cold and kept on working.

Bunkhouse scene from camp of Ole Emerson, near Cable, about 1905. The men are seated on the "deacon" seat, a bench hewed from a single pine or basswood log. The men wear "stag" pants, cut off below the knee, and the one in center has calked boots, probably a top loader on the landing. Straw filters between the outer frame and slats of the bunks, and wet socks hang out to dry. The men seem relaxed, for this is probably Sunday, and a camerman has come to shoot a picture with flash powder equipment. All the men apparently wanted to be photographed with their felt hats on, even the man in the bunk at right.

In the woods, the men were at work at the crack of dawn and they continued to work as long as they could see, at least a twelve-hour day. In the sawmills of Michigan and Wisconsin, the twelve-hour day was standard until 1881, when employees of the Muskegon Booming Association went on strike for a ten-hour day. The big lumber companies broke the strike in 1882, but four years later the ten-hour day was accepted throughout Michigan and spread to Wisconsin and Minnesota. But in the woods, it was still daylight to dark.

The author of the article on "Life Among the Loggers" in *Harpers* (1860), says red flannel shirts were the most widely used by the shanty-boys in Maine, and on their feet they wore "good boot moccasins." In 1860, raftsmen on the Wisconsin River also favored the red shirt, according to a story in the Wood County *Republican*. But the red-shirt fad died out. Later photographs taken of crews in the woods and on the river drives, although not in color, suggest that the men wore just about any kind of color or checked shirt. Light mackinaws or wool jackets with cross stripes were worn over the shirt. When the men warmed up, the jackets came off.

As to the "good boot moccasins," these were a type of footwear difficult to imagine today. They were without heels, and only slightly better than an Indian moccasin. In Wisconsin, the descendant of the boot moccasin was the "shoe pack," a leather or rubber shoe without a heel, ankle high, which was packed with wool socks. It was used mostly by teamsters "on the haul," not by the sawyers and swampers. Military-type leather boots, with high heels, were still in vogue in the 1880s, but then, gradually, the change was made to rubber boots, that is, a shoe of rubber with leather tops. For the men on the log drive, leather boots with studs or nails in the sole, called "calked boots," replaced the high leather boots seen on photographs taken by Bennett in 1887.

Equally popular was the quarter boot, a rubber shoe without a top that fit snugly below the ankle and was worn with at least two pairs of homespun wool socks, knee high.

Pictures of woods crews from the 1870s and onward reveal a wide variety of hats, mostly felt, as well as caps with ear mufflers. But the man who could go all winter without wearing anything over his ears was considered a "tough guy." The result was that some men, whose circulation was not as good as others, came out of the woods in spring with "cauliflower ears"—too many frostbites.

The work ethic was accepted in the logging camps of the 19th century just as it was understood on the wheat fields of Dakota Territory. But in every crew, whether in the logging woods or wheat field, there was a hustler who set the pace for the others. In one camp north of Rosholt there were two sawyers working for Mork & Jacobson, who cut more than any other team of sawyers for miles around. They were nicknamed the "Gold Dust Twins," after a strong soap powder of the period. For there was something heroic about men who set the pace for others, and they were looked up to. The hustlers were encouraged by the foremen and flattered just enough to keep them beating their brains out to excel and, in a way, to dominate the others.

But in addition to the work ethic and the hustlers, there was also the "tough guy" ethnic which prevailed in all camps. This was encouraged by the early Yankees and adopted by the immigrants as soon as they became part of the American Dream. No one stood around in the woods, for example, and complained about the cold. If he did, he was a "softy" who "couldn't take it." It was permissible, under certain circumstances, as on the landing where the wind was fierce, for a teamster to wear a scarf wrapped around his head to cover his ears, but he still had to be wearing a felt hat. Rarely did anyone wear a stocking cap with tassle, the kind once worn by the French fur traders and forest rangers.

There were warm dogskin gloves made with high cuffs, but these were for drivers who wanted to show off their young gelding behind a brand new cutter, bells a-ringing on the traces. But no logging teamster ever wore gloves, dogskin or otherwise. He wore wool mittens with leather mitt outside like the modern "chopper" used in the woods and by ice fishermen. Swampers and sawyers all used mittens.

For underwear, one piece "long handles" made

Cooks at a camp for Tigerton Lumber Company near Summit Lake, 1910. Left, Mrs. Charles Kannenberg, Lily and Lind Schneider.

of heavy wool were the most common.

One of the striking features, however, of the photographs taken of camp crews from the mid-1880s into the early 1900s, is the wide variety of work clothes. Most men came into the woods with their "old clothes," no matter what color or style. Overalls, however, were not used much before the 1880s and seldom in the logging woods. Farmers wore overalls, not lumberjacks, even if they came from the farm. But after 1910, overalls were acceptable, too.

There was little illness in camp. More than one lumberman, writing his memoirs, has remarked on this. Accidents were common, but that was something else. If a man broke his leg, or chopped the instep of his foot with an axe, he was treated by a doctor—if one was available—and allowed to sit around in camp for a few days, but then, if he could sit on a load of logs and drive a team, he was encouraged to do so; anything to get him on his feet and back to work. If a leg or arm had to be amputated, there was no compensation. Insurance policies for the shanty-boys were unheard of until the late 1880s when a Detroit newspaperman introduced the idea and it was taken up throughout the Lake States. Agents of insurance companies or agents of a local hospital, went to the sawmills and camps and sold "tickets" costing around $5 per year which entitled the holder to hospitalization, but not to any benefits beyond that. Dr. Michael Ravn had a hospital at Merrill in the early 1900s which catered to lumberjacks in Lincoln County.

In most photographs of camps in the woods there is no evidence of an outhouse. A photo taken of a Moore & Galloway camp east of Pike Lake (Marathon County) in 1891 appears to have an outhouse, but very small and probably reserved for the foreman and cooks. The teamsters probably used the barn or stable, the others went into the woods or used the camp trench which was ten to twenty feet long and had a stringer, supported by suitable framework, over it for the men to sit on. In some camps a leanto was built over the trench. There were no flies after November 15 and before April 1. Dry leaves or an old Sears & Roebuck catalog were the chief source of toilet tissue.

Very few of the men in camp ever smoked a cigar. The foreman might smoke a cigar, but pipe smoking was the most common among all ranks, with *Standard* and *Peerless* the most popular brands of smoking tobacco. Chewing tobacco and snuff, better known by the Norwegian word "*snus*" (pronounced snoose), were also popular. In fact, the word *snus* became a byword for anything strong. If a man was trying to lift something heavy and could not, he was urged by his companions to apply a little more *snus*.

After the railroads penetrated the northern Pinery, most camps of any size had a "wanigan," a small store that catered to the crew, selling necessities such as socks, shirts, mittens, thread and needle, and tobacco. It was usually located in the foreman's quarters, a specially built log hut or frame shanty. If the customer did not have cash, which was usually the case, he could charge his purchase and settle up at the end of the season. The wanigan was usually run by the bookkeeper, or the foreman, or the wife of the foreman.

The unidentified man on the dogsled was probably an agent for hospitalization. He is shown here crossing Deer Lake in Sawyer County.

Below, two tickets used by lumberjacks for admission to hospitals when injured or ill. J.J. Synnott, who gives his address as Wausau, probably purchased the one to Dr. Ravn's Hospital first, and later, as he moved north, the other one which admitted him to hospitals in Ashland and Duluth, and, points west and east. The ticket made out in 1904 shows that Synnott paid $10. Most tickets in the 1880s and 1890s cost $5. Agents of the hospitals traveled among the logging camps and mills and sold the tickets.

When Ordering Medicine, Address DR. RAVN'S HOSPITAL, Merrill, Wis.

No. 1936 **DR. RAVN'S HOSPITAL,** **$10.00**

708 SECOND STREET,

MERRILL, WISCONSIN.

Dated at Merrill this 1st day of Feb. 1904

This certifies that Mr. J. J. Synnott

Is entitled upon presentation of this Certificate, to admission, medical and surgical treatment in DR. RAVN'S HOSPITAL, MERRILL, WIS., at any time during one year from the date of this Certificate, in consequence of wounds, injuries or sickness hereafter received or contracted and which shall disable him from the performance of manual labor.

Subject to the conditions endorsed hereon which are hereby agreed to and made a part of this contract and signed by him.

The holder of this Certificate is entitled to medicine at the Hospital Dispensary whenever needed, or it will be sent free to any part of the United States or Canada. ☞ Also free Baths, including Turkish, Russian, Plunge and Shower.

When badly injured or very sick ~~TELEGRAPH ahead and ambulance~~ will meet you at the train.

By C C Smith Agent. **DR. RAVN'S HOSPITAL.**

Any Alteration Makes This Ticket Void. Not Transferable.

CONDITIONS.

The following are excluded from the benefits under the provisions of this contract, viz:

INSANITY, CHRONIC AND CONTAGIOUS DISEASES.

This Certificate is not transferable, and all its benefits terminate at the expiration of one year from the time of its issue.

This Certificate holder agrees to comply with the rules and regulations of the Hospital. Veneral diseases treated on this ticket at this hospital only.

HIS J. J. Synnott Signature. MARK

Home Address Wausau Wis.

When ordering medicine write plainly the number of certificate, your full name, post office and express office address. When medicine ordered is not received promptly notify me at once.

EXCHANGES—Alexian Bros' Hospital, Oshkosh, Wis., Marinette & Menomonie Hospital, Marinette, Jamestown City Hospital, Jamestown, N. D., Alexian Bros' Hospital, Chicago, Ill., M[redacted] Paddock Memorial Hospital, Tacoma, Wash., St. Luke's Hospital, Denver, Col., D[redacted]. St. Alexius Hospital, Bismark, N. D.

Dodd's Hospital, Ashland, Wis.; Red Cross Hospital, Duluth, Minn.

The company town

Some logging camps grew into small villages composed entirely of people employed by the lumber company, such as the Rust-Owen Lumber Company, which purchased thousands of acres of timber land in Bayfield County and built a big sawmill at a place later to be called "Drummond." In addition, the company built eighty-two houses for its employees on a rental basis, and a general store, and arranged to establish a post office. This was a "company town." Other well-known company towns include Hayward in the 1880s and 1890s, Heineman Lumber Lumber, northeast of Merrill in the first decade of the 20th century, the Goodman Lumber Company at Goodman, and the Shell Lake Lumber Company in Washburn County. A small company town was also located around the turn of the century at a place called "State Line" which later became Land O'Lakes.

One who remembers the sawmill days in Drummond is Mrs. Dolores Rifelman (Green Bay). She said the sawmill, and four houses where company officials lived, had running water, but everyone else in town had to walk to a town pump. Some people had a block to walk, others four blocks. She remembers having to wait in line for her turn at the pump. Many of the employees were Norwegian newcomers who had brought along their wooden shoulder yokes to America, or had them made after they got to Drummond so that they might carry two pails of water from the pump instead of one.

At home, Mrs. Rifelman said, her family had extra barrels to store water in summer—rain water—but it was necessary to make several trips to the town pump in winter to keep up with the demand, especially on "wash day."

These are the men who operated one of the biggest saw mills in Wisconsin, built general stores where their employees could trade, and made and sold coins and currency to be used at the company stores. They are the directors and stockholders of Knapp, Stout & Company of Menomonie holding a board meeting, circa 1900. Clockwise, seated, W.W. Cassidy, J.E. Stout, J.H. Douglas, H.E. Knapp, L.S. Tainter, F.D. Stout, and H.L. Stout. Standing, left, F.E. Wilson, J.H. Knapp, and T.B. Wilson.

Company-owned store of Hackley, Phelps & Bonnell Lumber Company, 1908, at Hackley, Wisconsin, today known as Phelps. Trade tokens, noted on following page, were accepted here. The overhead wires carried tray with cash back to cashier who took money and sent change back to clerk. William Phelps stands in front of counter, second from right. B.O. Irvin, at right behind counter, was store manager. Jerry Holland, representing U.S. Post Office and Indian Agency, stands in rear at left.

There was no electricity at Drummond until 1929, although the sawmill was generating its own electricity for many years. There were two street lights, one at the town pump and one a block down the street.

The general store was company-owned, and employees of the company could buy their provisions with coupons issued in denominations of $1., 50¢, 25¢, 5¢ and 1¢. A book of coupons was deducted from the wages of the employee. There was no competing store, and prices were rather high.

Three whistles from the mill alerted the workmen, one at five o'clock in the morning to wake up the barn boss and teamsters, one at six o'clock to wake up the sawmill and logging crews, and one at seven o'clock to start working. There was also a whistle at noon when the men, who did not carry a lunch pail, hurried home to get a bite to eat, but they had to be back on the job at one o'clock when the whistle blew. Coffee breaks were unheard of. At six o'clock the last whistle blew and the big band saws, and circular saws, and pulleys and belts and shafts came to a standstill.

The company also operated a large boarding house for those who had no family ties, and a hotel both for employees and transients. After the mill closed in 1931, renters who wanted to remain in Drummond bought the company houses. The land logged off by Rust-Owen is today part of the Chequamegon National Forest, which has already grown back to great trees that are being cut on a selective basis.

Gerald Johnson (Wisconsin Rapids), the leading collector of lumberjack tokens in Wisconsin, says there were at least twenty-two lumber companies in the state which sold paper notes and tokens to their employees. The paper notes issued by Knapp, Stout & Company were made by engravers in Chicago and resembled United States currency. Their paper denominations ran from five cents to five dollars, and their coins also from five cents to five dollars. Knapp, Stout & Company maintained three larger general stores, one at Menomonie, one at Prairie Farm (Barron County), and a third at Rice Lake.

The Wisconsin lumber companies that used paper and metal script tokens were Bundy Lumber Company (Lincoln County), the Chas. W. Fish Lumber Company, and Fish-Johnson Lumber Company, both of Elcho, the Vilas County Lumber Company, Fosterville, Goodman Lumber Company, Goodman (Marinette County), Yawkey Lumber Company, and Yawkey-Bissell, both of Hazelhurst, Larson Lumber Company, Jeffries (a name which replaced Bundy), F.L. Nichols Lumber Company, La Crosse, Connor Lumber and Land Company, Laona, White River Lumber Company, Mason (Bayfield County), H. Marcus and Sons, Muscoda (Grant County), Hackley-Phelps-Bonnel Company, and C.M. Christiansen Company, both of Phelps (Forest County), Ruby Lumber Company, Ruby (Chippewa County), R. Connor Company, Stratford, Badger Lumber Company, Tigerton, Mason-Donaldson Lumber Company, State Line (today Land O'Lakes), Vilas County Lumber Company, Winegar, Turtle Lake Lumber Company, Winchester (Vilas County), and Wolf Creek Lumber Company, Wolf Creek (Polk County).

There might appear to be a duplication in the Vilas County Lumber Company, one at Fosterville and one at Winegar, but tokens have been found for both because Winegar was a name that replaced Fosterville in 1910.

In its infancy, White Lake, Wisconsin, was a company town, dominating by the Yawkey-Bissell Lumber Company. Right: a row of family houses, all alike, built for employees with families.

Coins and currency

KNAPP, STOUT & COMPANY

RUBY LUMBER COMPANY

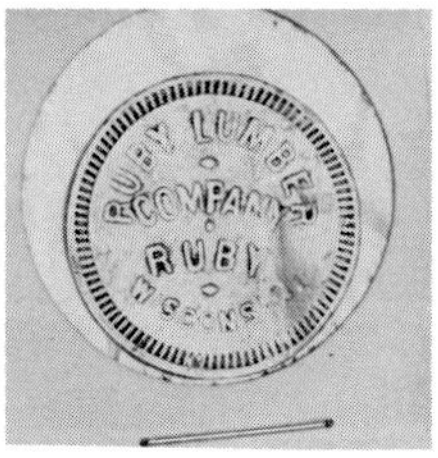

GOODMAN LUMBER COMPANY

WOLF CREEK LUMBER COMPANY

CONNOR LAND AND LUMBER COMPANY

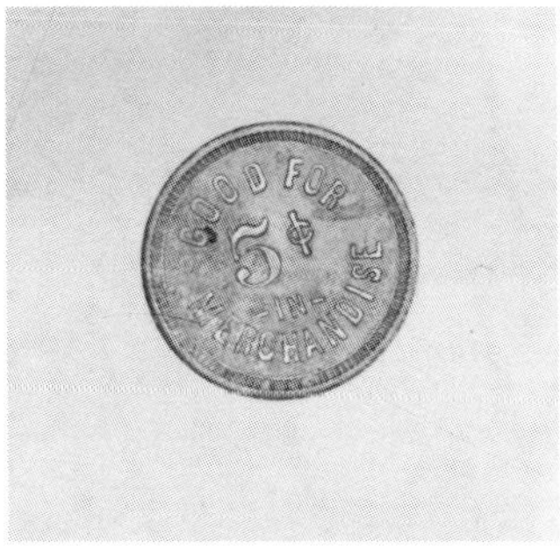

WHITE RIVER LUMBER COMPANY

C.M. CHRISTIANSEN COMPANY

MASON-DONALDSON LUMBER COMPANY

YAWKEE-BISSELL LUMBER COMPANY

VILAS COUNTY LUMBER COMPANY

An Indian lumberjack, probably a Chippewa, wears several ornamental buttons on his cap in a picture taken in January in 1892, probably on Red Cedar River for Knapp, Stout & Company of Menomonie. A picture taken a decade later of a log drive on the Rib River has two Indians on it, both with the same kind of buttons. This particular picture is the only one available from this early period showing the use of the corner-bind chain on a load of logs (left).

A couple goes for a ride in a cutter, probably near Rib Lake. The woman is well bundled. If a long drive was planned, it was customary to have a charcoal burner, or a piece of heated soapstone on the floor of the cutter to keep the feet warm.

These three pictures are, in subject matter, duplicates of several others found in this chapter. What distinguishes them from the others is their age and their shape. All were taken on January 21 and 22, 1892 by a cameraman using an Eastman Kodak that shot a picture with a round frame, as described on page 51. The ones shown here are also slightly enlarged from the originals. The written inscription (right) says "Dinner in woods, Ritchie's" meaning, no doubt, Ritchie's camp. Below: lunch sled at C. Johnson's camp," and at lower right, "Helliker's Camp" on east side of Big Chetek Lake, south view. The snow on the roof suggests where the dingle is, dividing the sleeping quarters from the kitchen area. All three pictures are probably associated with Knapp, Stout & Company of Menomonie.

Postcard picture (enlarged) of White River Lumber Company sawmill at Mason (south of Iron River), about 1910. White River company was subsidiary of Hines Lumber Company of Chicago. Sawmill shipped out products on Chicago, St. Paul, Minneapolis & Omaha railway, abandoned 1978. Shack at left in picture was for section handcar.

Blacksmith shop at Charles Kennenberg camp near Summit Lake, for Tigerton Lumber Company. Fred Volmer, cookee (right) holds hand on runner made by blacksmith for bob sled on company's steam hauler. A skidding tong hangs on wall of shanty behind the runner. Horseshoe above the door wards off local trolls.

C.J. Ruppenthal (Tigerton) took this picture of a blacksmith at work for Hatton Lumber Company's Camp 3 located south of Elderon. In the corner stand stocks waiting to be mounted with hooks and bell-toes. The lever above the raised hammer was used to pump the bellows (lower left) for the forge seen behind the blacksmith. Spare hooks for eveners, chains and clevis hangs on wall at right, and at left, pretty girls adorn a calendar.

"Car camps, or shanties built on railroad flatcars were used to house and provide kitchen facilities for crews working for the Vilas County Lumber Company at "Winegar," now a ghost town near Presque Isle. The picture may have been taken in 1913, and it shows that this company has been logging in summer. Perhaps that is why it was feasible to use these box cars for quarters, otherwise they would have been too cold in winter.

The picture above of a dining car in a "car camp" gives the location as "Fosterville," a name which was changed in 1910 to "Winegar." The crockery on the table seems to be enamelware, the most common in the logging woods all over Wisconsin. The men who sat at the one-benched table at the right had to reach for the corn syrup, etc.

Left, S.D. Brundage, who stands at the left in the picture, built this shelter to sit in while waiting for the next load of logs to come by. It was his duty to keep hay in the sleigh tracks on a nearby hill to slow up the runners on the downhill run. He also kept the sleigh tracks on the level clear of debris and horse manure. Men who tended to these chores were called "road monkeys," "hayman on the hill," or "sparrows."

A blacksmith shoes an ox in the open. An ox had to be trussed up with belly band, hind feet barely touching the ground, because loss of footing on one leg unbalanced it.

William Stuplish (left) and Edward Page at pump, probably in camp near Minocqua. Stuplich is chewing on a toothpick or match, a common sight after a meal in days gone by. He was probably the camp clerk. Page was a jobbing foreman and spent several years working for Hatten around Elderon-Galloway before moving farther north.

The skidding tong

A small step forward in the technique of skidding logs out of the forest came with the invention of the "skidding tong." This was really not an invention of principle, but an improvement on the common ice tong. In the early period of logging in Maine, a wedge was driven into the log, and a chain, hooked to the wedge, went forward to the evener behind the horses or oxen, and the log was skidded out to a tote road. Out of the need for a better method to skid a log a short distance, and to avoid damage to the log caused by the wedges, was born the skidding tong. It was hooked into the end of a log like an ice tong into a cake of ice, and the harder the horses pulled, the more firmly the log was held. It should be emphasized that both the wedge and the skidding tong were used only for short distance skidding out to a tote road. If any longer distance was involved, one end of the log was rolled onto a travois or go-devil, and in some camps, on a "train" or "dray" which was pulled on the ground like a stoneboat.

Perhaps the earliest picture ever taken of a skidding tong in the Lake States appears in a photograph taken near Stillwater, Minnesota, about 1880, where oxen are skidding a log with skidding tong. No reference to a skidding tong appears in any of the articles on logging in *Harpers* from 1851 to 1868.

It was common to "ross" the end of a log about to be skidded out. The bark on the butt was chipped around the circumference and chamfered so as to offer less resistance to underbrush, roots, and snow. If the log was extra large, and only one team of horses available for skidding, the log might be rossed the entire length on one side.

In an article appearing in *Harpers* (March 1860) entitled "Life Among the Loggers," one of the engravings shows three yoke of oxen skidding a big log. The log has been raised on one end, apparently onto a travois. The oxen, knee-deep in snow, are not pulling on the log, but a chain attached to the front end of the travois. According to stories handed down from the pioneers of Portage County, oxen often went blind from straining too hard, and, as an antidote, some pioneers allegedly poured melted butter into the ears of the blind oxen because they thought it would cure the blindness.

A dray for skidding logs preserved at the Menominee Logging Camp in Keshena. This type of dray could be used to skid several small logs or one or two large ones. The late Joseph Gruna (Rosholt) once recalled that he had seen this type of dray in a camp for the Wright Brothers Lumber Company of Merrill in 1908. The skidding tong seen here has nothing to do with the dray but has been hung for display.

Rivers for transportation

More than a hundred years ago there were no roads into the great Pinery of northern Wisconsin, and there were no railroads. There were only rivers by which the lumbermen could get logs to a sawmill, and from the sawmill there were rivers by which they could get their lumber to market.

The larger rivers of Wisconsin had long been used as routes of commerce and travel by the Indians, by explorers, by missionaries, and by military people, but the small rivers remained unused because they were choked with windfalls and brush accumulated over the centuries. But big or small, the rivers could be used, under the hand of man, to serve his needs. They were rivers of life.

For example, when Moore & Galloway Lumber Company of Fond du Lac began logging in Waupaca and Marathon counties in the 1870s, it was possible for the company to get its logs out of the woods by floating them down the Little Wolf River into the big Wolf and on to Lake Poygan where they were sorted at "Bay Boom" and forwarded on to Fond du Lac via the Fox River and Lake Winnebago.

But before the company could use the Little Wolf, it had to send clean-up crews to clear the river of windfalls, stumps and "rampikes" (half dead trees), that had fallen into the river. It had to remove large rocks from the middle, which might impede navigation, and finally, it had to build a network of dams extending all the way from Marathon County through Portage and Waupaca counties, to the big Wolf in Winnebago County.

To clear the river of rocks, men used black blasting powder to break them up. Oxen, pulling on a windlass that went with four sweeps, moved the larger pieces of rock ashore or into a hole in the river where they would not block oncoming logs. When logs came downriver, any snag, rock, or impediment could become a trap and if one log got hung up, it quickly started a chain reaction that led to a pile of logs known as a "jam."

The banks of the rivers had to be cleared of brush, too, because the men who ran the logs, that is, guided them in their course downriver, had to have paths, called "gig paths," to follow up and down the river on foot. On the larger rivers, like the Wolf and Chippewa, two men in a bateau, using oars and pike poles, moved from one bank of the stream to the other carrying the drivers to any spot where there was a threat of a jam. (The word *bateau* comes from the French for boat, and the river men used the French word for it because the boat was brought to Canada by the French, and to the Lake States by the French-Canadians.)

The rafting of lumber and the driving of logs on the rivers of Wisconsin was made legal by the adoption of the Northwest Ordinance on July 13, 1787, more than a year before the Constitution of the United States was adopted. Article 4 of the Ordinance provides that all "navigable waters leading into the Mississippi and St. Lawrence, and the carrying places between the same, shall be common highways, and forever free . . . without tax, impost or duty therefore."

These provisions were incorporated into an act admitting Wisconsin into the union of the states on March 4, 1849. Article 34 of the statutes (1849) states that "all rivers and streams of water in this state, in all places where the same have been meandered [surveyed] . . . are hereby declared navigable to such an extent that no dam, bridge, or other obstruction may be made in or over the same, without the permission of the legislature."

On the upper Wolf River in Langlade and Forest counties, a network of dams extending all the way from Pine Lake to and through the Menominee Indian Reservation, had to be built to make this river suitable for log drives. The first dams were built on the lower rapids, beginning in 1868, at the Dells north of Keshena. Since the Wolf takes its rise in Pine Lake (Forest County), the altitude here is higher than it is at Keshena. It follows then that the river is flowing down a long hill or basin. But the slope is uneven. There are spots where the terrain drops off sharply and the pitch in the river is steep. At places such as these the river has loosened the gravel in the bed and left rocks and hardpan either exposed or submerged. These places are called "rapids" and where there is a continuous stretch of fast-running rapids, there is "white water," the same that all canoeists keep a lookout for because they know that by following the white water they are in the main current of the stream.

The men driving the logs down the river, called

"river pigs" in the lexicon of the river, often rode a log through dangerous stretches of white water, and anyone who could do this and not fall off and drown, was called a "white water man." Such a one was "Whitewater Ole" (Horne) on the *Chippeway* drive.

In order to overcome the rapids, that is, in order to raise the level of the water above the rocky stretches or shallows, it was necessary to build dams to hold back the water in reservoirs called ponds or flowages. According to Randall E. Rohe, who has made a special study of the subject, the dams were built of cribwork, that is, square boxes of logs filled with small rocks and gravel. On the upstream side of the cribwork, from the crest of the dam to the bed of the river, a row of closely placed planks of split posts, called "toe piling," was weighed down with earth to keep the water from seeping through or under the dam.

For log driving purposes, most dams were built with sluiceways, or gates. A wall of timbers, laid along both banks of the river at an angle of some twenty-five degrees to the main sluice gate, formed a chute that made it easier to guide the logs into the sluice gate for the run through the dam. Below the sluice gate, an apron of heavy planking, or hewed timbers, was laid down to accept the pressure of the logs as they tumbled headlong through the gate. When these timbers or planks became weakened by age or seasonal change, a log could smash through the floor of the apron and stand on its head, creating a sudden barrier and almost surely a jam unless the men could lower the sluice gates in time and stop the drive.

Most gates on the "driving dams" were raised and lowered with winches of some kind. The gates slid up and down in slots on either fender like a window in a sash. These sluice gates as well as the waste gates were of uniform size, six or seven feet wide. Anything wider would have been difficult to lower and raise in a hurry.

Permitting the water to fill up a pond above its natural height in relation to a dam was called "gathering a head" (of water), or "catching a head." By releasing a "head," (i.e. surplus water) in one grand splash, the river below the dam was raised high enough to allow the logs to pass over a rapids or shallow stretch in the river. When all was ready, the sluice gate was opened and the logs were allowed to run through the gate "all to once." This was called "sluicing."

At Little Falls (Holcombe) the logs were sent through special gates called "sluice gates," or "rolls," which were about fifteen meters wide, supported by heavy pine trusses. On the first dam, which was built here in 1878, there were two gates at the left, or west, end (see photo on page 157), one for sluicing logs, the other for sluicing the wanigans or cook boats. Each crew on a beat had its own wanigan and when it arrived at a dam, two men usually rode the wanigan through the gate, clinging desperately to the long oars amidships, fore and aft.

The wanigans on the Chippewa were built in sections and before the dive through the dam, these sections were taken down and laid flat on the deck; the wanigans usually submerged a few seconds after passing over the apron of the dam into the deep water.

The first dam at Little Falls washed out on September 17, 1884, and a new one was completed in 1885. This had three gates, one for sluicing logs, one for wanigans, and a third and narrower one for sluicing the bateaux through. The business section of the new dam was built on a north-south axis in the river and the three gates were located on this axis, the main sluicing gate northern most, the wanigan gate in center, and bateau gate on the south, all separated only by the log cribs, about two by four meters, filled with rock and gravel which supported the main structure of the dam.

On the flowage side of the dam, that is, the upper side, the gates at Little Falls were not built in horizontal manner like the gates found on the dams of the Wolf and smaller rivers of Wisconsin. At Little Falls the gates were actually a phalanx of heavy planks, each about five meters long, which stood upright but leaning against the dam structure at an angle which the pressure of the water in the flowage held securely in place. These were called "splash planks," and when it was necessary to build up a head of water for sluicing, the planks were lowered behind the gate with a rope on a block attached to the top of an A-frame about four meters high. A hole was augered into one end of the splash plank and the rope from the block was passed through the hole and secured. Two men on a winch then wound up the slack and lifted the splash plank. A third man manually guided the plank into position. When a head of water had been built up, and the crew was ready to begin sluicing, the splash planks were pulled out with the same device and laid at the side of the plank road that ran the length of the dam and was wide enough for a team of horses.

The dam at Little Falls was built and controlled by Chippewa Lumber & Boom Company, a Weyerhaueser subsidiary, and all instructions to begin sluicing or to close the gates, came by courier from the head office in Chippewa Falls where the company's big sawmill was located.

There were, on many smaller streams, "roll" or "rolling" dams built without sluice gates. After the water reached the top of the dam, the logs simply rolled over the top and fell onto the apron below and gradually got carried downstream by the flow of the current. In this case the logs did not all come down the river at the same time, and there was time for them to flop over the dam without being crowded from the rear. On some streams these were called "splash dams" or "flood dams."

When the logs came through the several dams on a river, there were usually logs left behind that either "saddlebagged" on a rock in the center, or got stranded in a slough or on a sandbar. An accumulation of logs caught on a rock in the middle of the river was called a "center," and logs accumulated along shore were called "wings." To dislodge a "center" was often hazardous, but the cause of the blockade was usually one log, and when that was disengaged, the others tumbled downriver by the force of the current.

There were two crews on a log drive, the "rear crew" of twenty to forty men, and the "jam crew" (called the "beat crew" on the Chippewa) of ten to fifteen men. It was the business of the rear crew to pull, push, and remove the logs that became stranded along shore and formed "wings." This was also called "sacking." However, any log that was being rolled, or canted into the water was also being "sacked" into it.

Liberties taken with the facts of life in the woods and on the rivers were not confined to Maine or the Lake States. The illustration above is reproduced from an engraving appearing in a German publication, *Gartenlaude*, published in 1860. A translation of the German caption accompanying the illustration reads: "Log rafting in the Black Forest."

The jam or beat crew went ahead with the drive, keeping the logs from becoming jammed at any point, but not bothering about the logs that floated ashore. This crew also did not work in one place as a team, but spread out and took up positions where experience had shown there might be a jam.

Dam builders, with the approval of the state legislature, were permitted to charge a toll for logs driven or "sluiced" through their dam. For example, Chippewa Lumber & Boom Company, which controlled the famed Cameron Dam on the Thornapple River, was permitted to charge ten cents a thousand (board feet) for logs sluiced through their dam. But no toll could be levied by an independent lumberman or farmer without a franchise from the state.

As fanciful as the illustration in the German publication, the two on this page appeared in American magazines, the one at the left in an article on "Life Among the Loggers," *Harpers* March 1860, captioned "The Jam." However, if the jam breaks, scratch two good men balanced in midair. In the illustration below, the source is not identified, but it may have appeared in the 1840s. It shows that the driving crew is using only hand spikes and axes to move the logs, while the man with his shoulder to the pole may be using a jam pole. The high leather boots worn by the crew were popular before and at least two decades after the **Civil War.**

This photograph was taken either on Tibbett's Brook or Rum River (Minnesota) in 1864 by a photographer from the Whitney Gallery in St. Paul. The spring drive is under way. Some of the crew may be using jam pikes, but at least four are using peavies, the one at the far left being especially noticeable. This is one of the earliest log driving pictures in the Lake States, and confirms the use, in 1864, of the peavey this far west from Maine where it was invented in 1858. The photograph seen here was used as a model for an engraving in *Harpers New Monthly Magazine* for March 1868 in an article on "The Minnesota Pineries," and captioned "Breaking a Jam." Actually there is no jam in the original photograph, only a "wing" that has formed on the bend of the river where logs have been caught in the brush and overhanging branches. The crew was employed by Elias Moses, a lumberman from St. Anthony, later known as Minneapolis. This picture is enlarged from the original which is eight and a half centimeters wide and seven and three quarter centimeters high.

The photograph above may be the earliest ever taken of a cook boat, or "wanigan," in the Lake States, another in a series taken on or near Rum River in 1864. This photograph was also used as a model for an engraving in the *Harpers* article mentioned on the preceding page, as seen at left. The wanigan is switched to the opposite bank of the river in the engraving, which was probably an error on the part of the printer who reversed it even as he did to three or four others in the same article in *Harpers*, leaving historians to wonder why some of the engravings follow the original photograph in direction and some do not. The wanigans of the period appear to be skiffs with a piece of canvas over the stern to protect the food, but the stove and pipe are in the center. A table has been set up ashore for the crew to eat at, standing.

A wanigan probably on the Big Eau Pleine (Marathon County), about 1900. The men have had their lunch ashore and one of them is finishing a last cup of coffee.

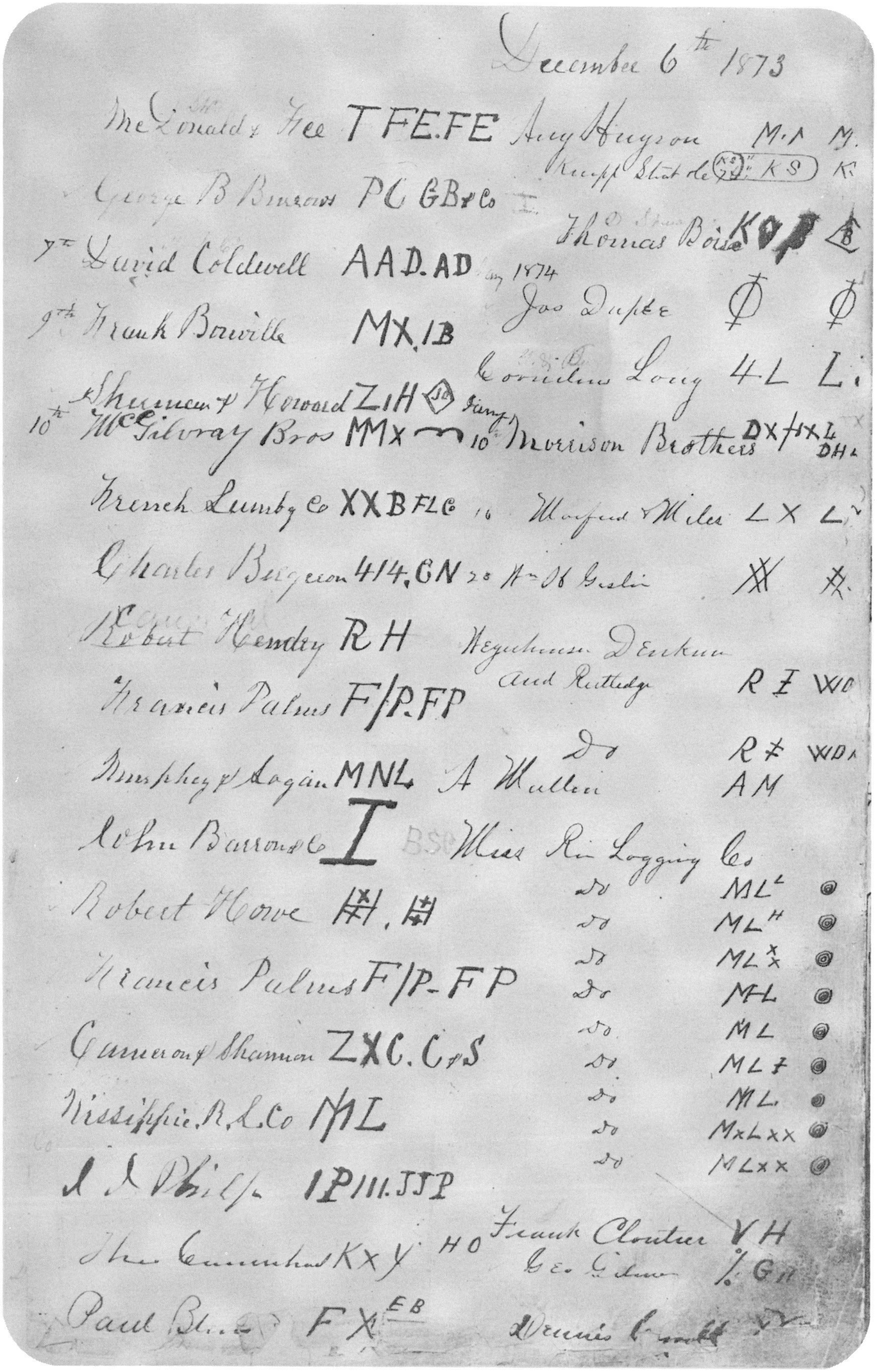

The abbreviated lettering in the above relates to log marks used by lumber companies to identify their logs. The dating on the document is December 6, 1873, and all the names appear to be related to the Chippewa River Valley. The names at the left may be jobbers, that is, contractors who were cutting timber for the companies listed at the right of the markings, including Knapp, Stout & Company, Weyerhaeuser, Denkum and Rutledge, and Mississippi River Logging Company, the last name also a Weyerhaeuser organization.

Two bateaux approaching Cedar Falls on Flambeau River.

An imprint made on a postcard from the original of this picture reads: "Ready for sluicing at Clam Falls." The village of Clam Falls was once located northwest of Shell Lake on Clam River in Polk County. The picture was taken in 1894, and standing fourth from the right, leaning on a peavey, is Louie Blanchard who years later would be featured as a story teller in *The Lumberjack Frontier*, edited by Walker D. Wyman.

The Wolf River drives

When Philetus Sawyer, one of the leading lumbermen of Oshkosh, learned of the great timber reserves lying to the north of the Menominee Indian Reservation, he got a franchise to organize the "Keshena Improvement Company," and to build the first dam above Keshena at the Dells, in 1868. Gardner Dam, about fifteen kilometers above the Dells, was built the same year. It was probably named for George Gardner, a Stockbridge Indian who lived near White Lake. Here five men lost their lives. It had rained hard and a big head of water was threatening the dam. Eight men were called out in the middle of the night to raise the gates, and just as they began, the bulkheads caved in, sending them plunging into the wine-dark river. Four survived, one of them John Satterlee, a well-known Menominee and Civil War veteran.

The first drive on the lower Wolf was allegedly run in the spring of 1843 by Samuel Farnsworth of Green Bay, and his foreman, Charles Westcott, destination: Oshkosh. Obviously this was not a big drive. The logs were probably cut no farther north than the present Orihula tavern. The last drive on the lower Wolf passed Fremont either in 1910 or 1911, and when these logs were sorted at Bay Boom, an era covering fifty years of booming on Lake Poygan ended.

The direction taken by the log drivers was controlled by the geographic tables established in the last Ice Age and which created the great river systems in Wisconsin. The eastern watershed, made up of the Wolf River, the Fox and Menominee, the Peshtigo and Oconto, and lesser streams, drain into the St. Lawrence watershed which leads to the Atlantic ocean. Thus, an empty bottle tossed into Pine Lake in Forest County might enter the Wolf and float to its mouth on Lake Poygan, and from there to Lake Winnebago, back into the Fox to Green Bay and Lake Michigan, and to the Atlantic. A bottle tossed into Lac Vieux Desert, headwaters of the Wisconsin River, might conceivably float down the Wisconsin and Mississippi rivers into the Gulf of Mexico.

The first rapids below Gardner was called "Garfield" and the second, "Hanson." A kilometer or more below Hanson's lay "Gilmore's Mistake." This rapids actually lies on the north line of the

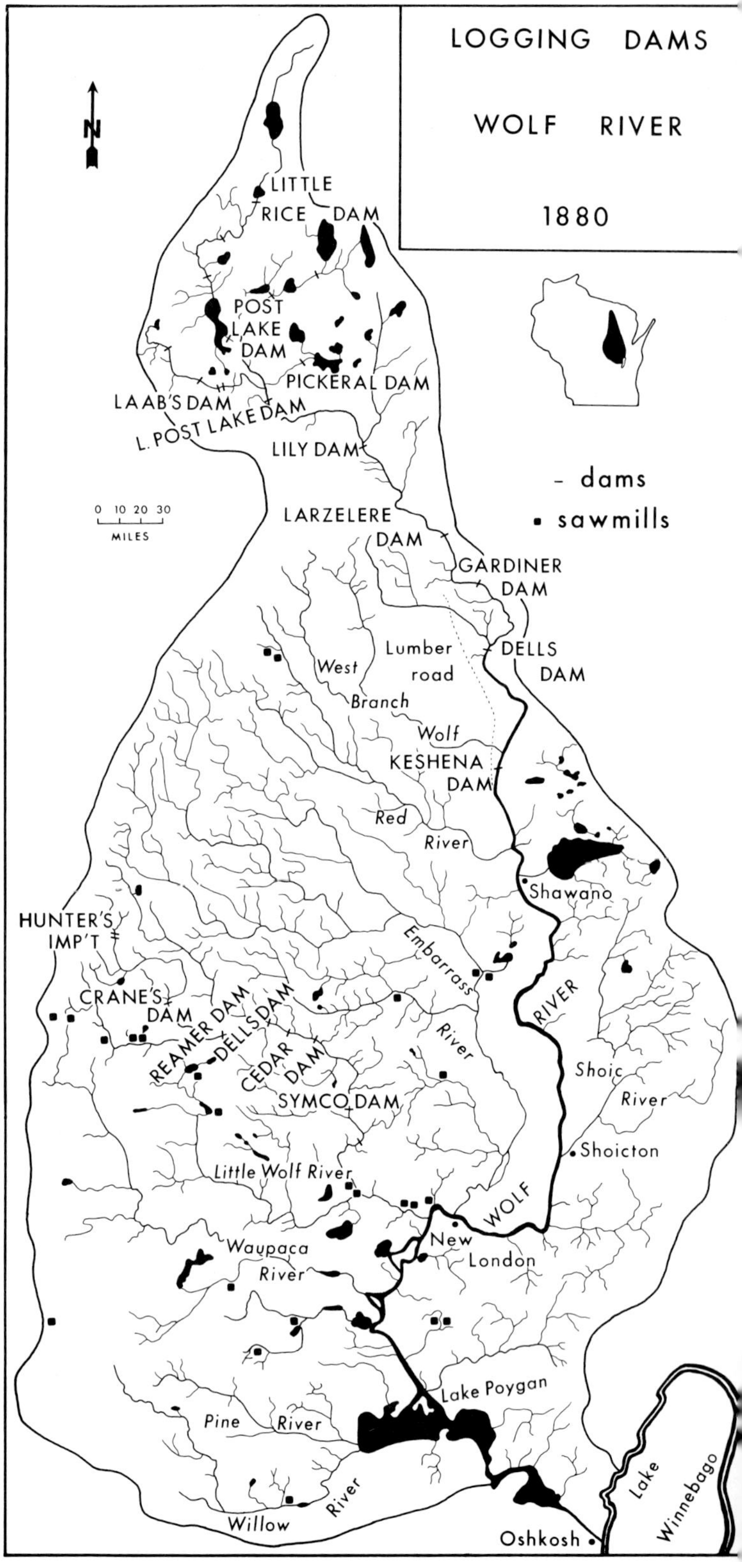

A map by Randall Rohe shows logging dams on Wolf River and also on Little Wolf in 1880.

Indian Reservation. According to legendary accounts, a party of timber cruisers, led by one John Gilmore, was cruising for timber on the Wolf River in the late 1850s, and when the crew reached the north line of the Reservation, the water seemed to boil right out of the ground like a huge spring. Actually, the river was choked by a curtain of trees, windfalls and brush. Gilmore assumed that the river originated in this boiling current and turned back. But a later party found that by going around the curtain of wilderness, the river was still there, flowing down from the north. This was "Gilmore's Mistake," a byword in the lexicon of the log drivers ever after. Kingsbury of Antigo took a picture of it, about 1900, although by that time the brush and windfalls had long since been removed.

The source of the Wolf River is Pine Lake in Forest County, and Dam No. 1 was located at the outlet of the lake. Farther down, at the outlet of Rice Lake, which the Wolf traverses, was Little Rice Dam, and farther on down, Pelican and Little Chute dams in Oneida County.

The Wolf swings east from Oneida County into Langlade County, thence southeasterly all the way to the Indian Reservation. After the river passes through Upper Post Lake it is joined by a small stream flowing north out of Lower Post Lake, and it then continues southeast to Pearson, Lily, and other towns. There was, and still is, a dam, south of Upper Post Lake, and another at Pearson called Lower Post Dam. The latter was actually built in 1874, and the one at Upper Post in 1878.

Meanwhile, the upper Post Dam, once made of timber bulkheads, brush and gravel, has been replaced by a concrete dam and this maintains the level of water to assure minimum erosion along the banks of Post Lake.

The next dam below Pearson was at Lily, and the next, the so-called Larzelere Dam (after Elton Larzelere) at Langlade. Gardner Dam lay below Langlade in Section 25, and below Gardner, several kilometers, the Dells Dam which stood at the head of the Dells on the Indian Reservation. The last driving dam stood at Keshena Falls. The dam at Shawano was a sawmill dam and built with a slide in the center for passage of logs.

Log drives of the Oshkosh and Neenah lumbermen were permitted to pass through the Indian Reservation, but they were not permitted to cut any timber on the reservation. The Indians cut their own timber and banked it along the river systems. In 1899, for example, the U.S. Indian Agent was advertising in the American *Lumberman* for bids on logs banked on the Wolf River within the Reservation. These logs were sold and the money turned over to the Indian jobbers.

Lack of rain was the main cause for most jams. Low water exposed the rocks and shallows in a river, which the logs could not float across without turning and twisting, and, in a matter of minutes, jamming up, one atop the other.

Too much rain could cause a river to flood and strand logs ashore, too. But a log jam could do the same thing. The jam, forming a wall against the river, would back up the water and raise the level above the banks, and logs still coming down the stream would then float ashore and into the sloughs and bayous.

One spring, in the 1880s, according to the late J.L. Whitehouse, a well-known river driver, "the Sherrry drive" was moving down the Wolf. (Henry Sherry was a lumberman from Neenah.) The logs jammed on the rocks at the foot of Tea Kettle falls and held the water back so that the logs, forced over the top of the jam, dropped down in front of it, creating a front sixty feet high. It took a crew of men two weeks to break this jam, but Whitehouse said this was in the days before dynamite was used.

After the logs passed through all the dams on the Wolf, including the sawmill dam at Shawano, the rest of the way to Bay Boom was in slow water, and there were no more dams to sluice through. A subscriber to the Wisconsin *Lumberman*, who had come from Stevens Point or Wausau, wrote an account of a trip he made down the Wolf River. It is dated Menasha, May 19, 1865 and his letter reads in part as follows;

> We steamed down the Wolf stopping a number of times to clear obstructions in the shape of log rafts floating down the stream. These were quite numerous. All along the banks, at every available point, were rafts of logs tied up, or booms filled with logs brought thus far on the drive. To one

Two men in a bateau running the Dells on Wolf River. Drivers at left and in distance stop to watch Kingsbury take picture.

accustomed to lumbering on the Wisconsin, the system of operating on this stream is somewhat strange . . . . When logs are once in the main river, it is almost equivalent to getting them to market. There are no obstructions, no 'flat' channels, and the lumbermen can 'bide their time,' and take advantage of the market. There is a responsible boom company, who attended to the division of logs, which are run into their booms promiscuously. After entering Lake Butte des Morts, and in many instances before, the logs are rafted, and run to the mills at Oshkosh or Fond du Lac.

"An inspection of the booms would afford considerable amusement to the Wisconsin Pinery Boys. They are built against piles instead of piers, and the stick [log] used for boom timbers are but little larger than our ordinary oar stems. They wouldn't stand the pressure of Ole Wisconse for twenty-four hours. Yet they are sufficient for all the purposes of lumbering, a vast amount of which is done on these streams.

"The amount of logs cut on the Wolf and its tributaries the past winter is about 130,000,000 [feet] but owing to lower water in the smaller tributaries, only about 80,000,000 will arrive at market the present season. In past years many mills have been built on the Upper Wolf and its tributaries, nearly all of which are now almost worthless, from the fact that the market at the lower towns is much better. And delivered at the mills of Oshkosh or Fond du Lac, there is no waste timber, even slabs and edgings are made available.

"The banks of the river at Oshkosh are lined with steam sawmills, all of which are in full operation. They manufacture immense quanties of lumber, and the comparatively small expense of manufacturing, the facilities for transportation, and the quick returns make the business pleasant and profitable. Whole freight trains are loaded for the Chicago market, and even logs are rolled upon platform cars and carried to market. And yet wiseacres pretend to say that lumber can be run cheaper from Wausau to Portage City over dams, rapids and sandbars than by rail around all those obstructions. It is sheer nonsense. Perhaps if some of those fellows had been stuck for three or four times on the Stevens Point dam, or run Conant Rapids with nothing but a 'dog oar' they would change their opinions . . . ."

The sum and substance of all this appears to be that the lumbermen on the Wolf River Valley were making more money than those in the Wisconsin River Valley because it was easier and cheaper to get the lumber to market by rail instead of rafting.

J.L. Whitehouse, mentioned earlier, was not only a river driver, but also a demolitions man. Said he:

> My regular station was at Twenty-Day Rapids about four miles downriver from Langlade. This rapids is nearly a mile long and considered one of the bad places on the river. I tended that mile alone 'til the rear [crew] came, and with the aid of dynamite didn't have a jam I couldn't handle . . . ."

Here only one man, supplied with dynamite, was "tending" the log drive as it came by. When he saw a jam developing he got over to it, presumably in a bateau or by riding a log, and inserted a long stick capped by a charge of dynamite, ignited the fuse, and retreated before it detonated.

Again, Whitehouse:

> We were camped at the Corn place and went via the Dells Dam. When we got there, there was no one at the dam, so we concluded there was a jam in the Dells a quarter mile below. We went there and found a jam, the sluicing crew and the two dam tenders. One had a charge of dynamite and a sixty-foot fuse, trying to get it in place from the rock above. I told Guy to have the charge pulled up and I'd blow the jam. I took a peavey and went and located that log that caused the jam on the far side. Then I tied the charge [of dynamite] to a pole, cut the fuse to three feet long, placed the charge, lighted the fuse and in three minutes the jam was gone . . . I had twenty-five jams in the Dells that year and never used but one charge on any one jam . . . .

At times like this the crews worked all day and most of the night for two or three days, with scarcely any sleep and little to eat. There was scant relief for a man who got hurt or sick either. Each crew had its own logs to run, and any delay in the progress of the drive at one dam or another could jeopardize the drive up, as well as down the river.

When there was only one consignment of logs to handle, one crew often followed it all the way to its destination. This often happened, for example, on the Little Wolf where a crew, starting out at the Upper Dam (Marathon County) continued all the way southeast to Bay Boom on Lake Poygan.

No sympathy was wasted by the foreman. Everyone was expected to accept his share of punishment from the river: the chill, the rain, the wet feet, and the lack of decent sleeping quarters. Thus, the ones who stayed on the job and did not quit after the first time they fell in or had to wade into the cold water to "sack" a stranded log, were the bravest of the brave, physically tough, mentally alert, and with ice in their veins.

An aprocryphal story is told about the crew that had put in an extra long day. The foreman came into the tent after supper and said, "Boys, we kept you at it pretty late tonight, but we'll make it up to you. Tomorrow morning we'll get an extra early start."

There were "dead man's bends" on many rivers. These were the legendary burying grounds where some log driver lost his footing and drowned and was buried on a knoll overlooking a bend in the river. Today no one remembers where these bends are, for time has erased these memories.

The largest dam on the Wolf River was at the Dells and this had four gates. Most dams on the smaller streams had two gates. The famed Cameron Dam on Thornapple River had three gates, and the Reamer Dam on the Little Wolf had three gates, all the rest were two gates.

Probably the last drive on the Wolf was run in early May 1924 when three million feet of hemlock arrived at Lily from Upper Post Lake. The logs were boomed at Lily, taken out of the river and shipped on the Wisconsin & Northern Railroad to the Yawkey-Bissell sawmill at White Lake. Three men handled the entire drive for Charlie McGee, the foreman, who a decade earlier had been woods boss for Moore & Galloway Lumber Company at Galloway.

The Brule-Menominee river system, which forms part of the boundary between Wisconsin and Upper Michigan, was another important log driving stream affecting Wisconsin. The lower end of the river, called the Menominee, empties into Green Bay at Marinette. The two rivers, fed by smaller streams large enough to float a log, made it possible for the lumbermen to harvest most of the pine in this region without the expense of laying down railroad spurs into the timber.

The timber in the Brule-Menominee basin seemed inexhaustible, and every year, for more than fifty years, the drives came down, either from camps along the main river, or from camps farther back on the feeder streams. In the first decade after logging began on the Menominee in the 1850s, there was no problem about log ownership because the drives were small and few between. But this situation changed in the period of the Civil War when greater demands for lumber brought larger crews into the woods and more logs into the river. As a result it became difficult to keep the log drives apart and disputes arose over log ownership.

In 1867, the Menominee River Boom Company was organized to bring order out of chaos, and a vast system of booms and gaps was built at Marinette-Menominee to hold the logs until they could be sorted, scaled and distributed to their owners.

The late Fred C. Burke, a local historian at Oconto interviewed by the author, believed that from 1867 to 1917, a total of ten billion, six hundred million board feet of lumber were floated down the Brule-Menominee. The peak year was probably reached in 1889 when 4,245,763 logs were harvested representing 642,137,318 board feet. Since each tree bucked up into three or four logs, it may be assumed that not less than a million pine trees were cut that year in the Brule-Menominee basin.

The widest dam on the Menominee River, called the Pemene, was built in 1888 at the head of Pemen Rapids. It had fourteen gates and two spillways.

In the 1870s, Holt & Balcom Lumber Company, and the Oconto Company, purchased timber rights on McCaslin Brook, a feeder stream of the Oconto River. Here dams were built to enable the lumber companies to get their timber to mill in Oconto. There were only two dams on the Oconto itself, but on the McCaslin, there were five. The first, near the mouth of the brook, was the Lynes dam, followed by the Sheridan, the Archibald, the Wheeler, and the Reservoir dams.

Kingsbury stood on top of "Big Jim Rollway" at Neopit and took this picture of logs swirling through a rapids of the Main West Branch in the Menominee Indian Reservation.

The two pictures above were taken at Big Jim Rollway on the Main West Branch at Neopit. This is not a log jam, merely a mass of logs that came loose prematurely and rolled into a heap. Kingsbury's caption in the upper picture explains that the rollway was 102 feet in *hight* (sic). In the lower photo a crew begins the task of breaking up the logs, probably using horses to pull the ropes.

Above: a crew is busy sluicing logs through Dells Dam on Wolf River in the north half of the Indian Reservation. Below: a close-up of the dam, the largest on the Wolf with four gates. In this picture one gate is open for sluicing.

An overnight tent for the log drivers on "Burned Shanty Rapids" on the Wolf. Here the logs have formed wings but the wings have been allowed to remain where they are to narrow the river and give greater force to the main current.

The rear crew is waiting for the flood, says a Kingsbury caption on this picture, meaning that the stage of water is down and the log drive has been suspended until a head of water has been built up in a dam farther upstream. When the sluice gates are opened, the flood will come down and these men will begin "sacking" stranded logs into the river. Standing up front in the picture appears to be an Indian driver, no doubt a Menominee.

Cedar Dam on Little West Branch of the Wolf had two gates, and two drivers with peavies (right) are watching the logs as they come through the sluiceway.

Near the north line of the Menominee Indian Reservation there was once a stretch of river known as "Gilmore's Mistake." Many years earlier, John Gilmore, leading a party of timber cruisers, thought this was the fountain head of the Wolf River. The woods and brush had completely clogged the river at this point and the water boiled up from the bottom like a great spring. Gilmore turned and went back, but a later party of cruisers went around the impenetrable jungle and found the river, still a long way from its source.

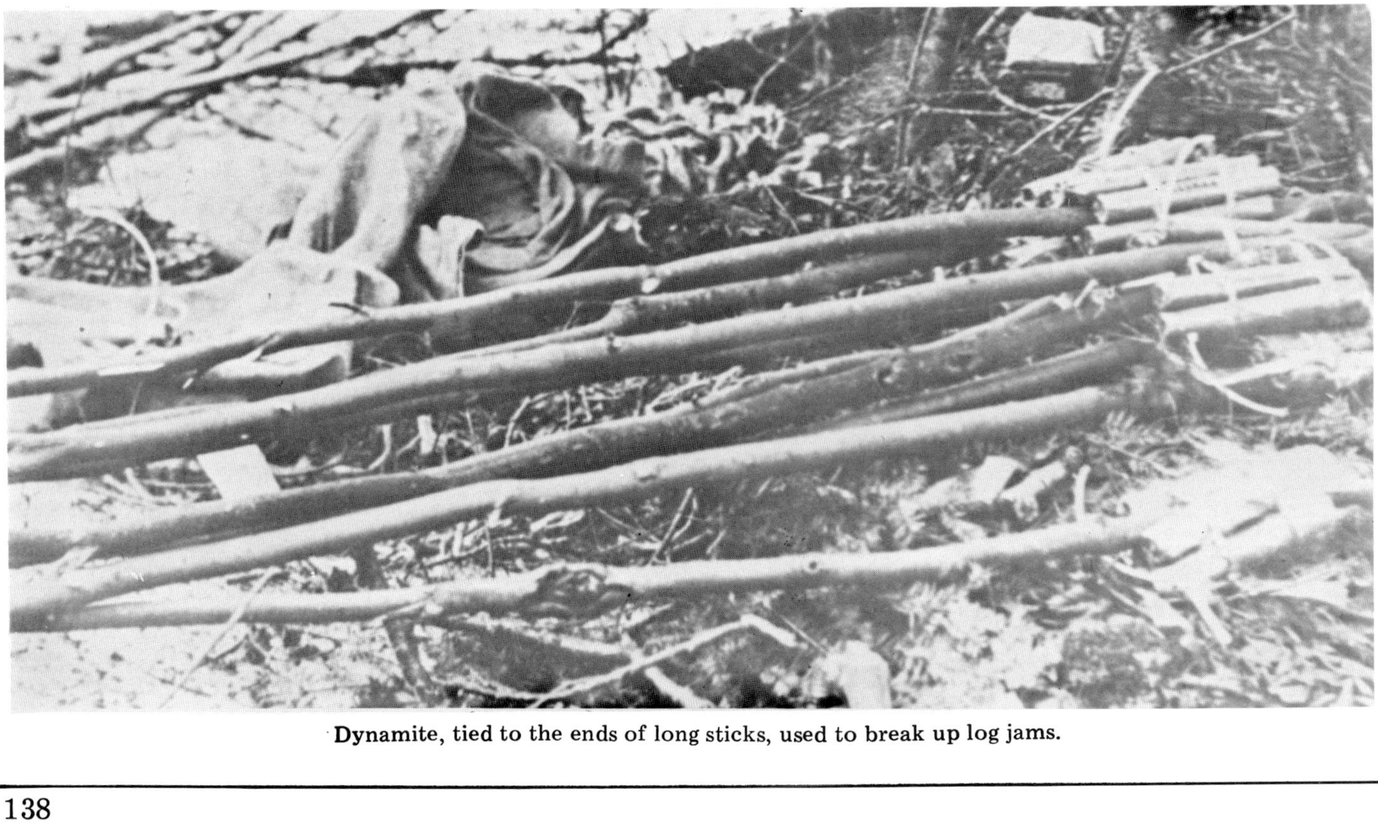

Dynamite, tied to the ends of long sticks, used to break up log jams.

Here dynamite is being used to break up a "center," or log jam over rocks in middle of river. Finding "key" log to loosen log jam was a myth. Most log jams were five to fifteen feet high and could extend upstream many rods or even miles. They were broken up with oxen or horses pulling on long chains or ropes, one log at a time, which often took days or even months to break up. Common sense dictated that whenever dynamite could be used, it would be used. Even setting the cap and fuse on a dynamite stick, as seen on opposite page, was a task for trained specialists, and most driving crews had at least one. Along the hillside (background) runs a thin line rising up the hillside through the trees. This is probably a "gig" path used by the crews when "gigging" back and forth along the river. Photo probably taken on Wolf River.

Drivers on the Little Wolf River in the Dells west of Big Falls (Waupaca County). Wm. R. Parks, a photographer from Iola, took this picture, probably in the spring of 1891. The Dells Dam was up the river a short distance to the left, and these logs, already sluiced through the dam, have formed wings. The crew must free the logs and keep them moving.

A caption at left, faintly visible, reads: "Sluicing at Big Falls, Wis." The picture, by an unknown photographer, was taken on the Little Wolf River in Waupaca County. Four men in the foreground and two at the left are keeping an **eye on the logs as they come through the dam, their peavies held ready for action in case a jam should develop.**

Below, an abandoned driving dam at Reamer Rips in the town of Harrison (Waupaca County) on Little Wolf River **becomes** a good trout hole a few years after the last drives **were made here,** circa 1900. This was a three-gate dam, but **the third bulkhead does not show clearly in the picture.**

Bay boom

The second largest booming works in Wisconsin was located on the lower Wolf River. Chapter 540 of Private and Local Laws for 1856 authorized Benjamin Brinkley and associates to maintain a boom between Gills Landing and Merton's Landing, the latter a name no longer used, but today associated with the community known as Orihula. It is uncertain just where this boom was located in relation to either place, but it patently outgrew its capacity for log storage and not long after, a new booming works was established at the mouth of the Wolf on Lake Poygan called "Bay Boom" or as "Boom Bay." This had a storage capacity of more than a hundred million board feet of timber.

Another boom was located on Boom Island, farther upstream, which had a capacity of three million feet; another at Merton's Landing with eight million, the Old River boom with five million, and the Synder boom with four million. These figures are based on an advertisement appearing in the Wisconsin *Lumberman* in 1874.

At Bay Boom a canal was dug from the main channel of the river across a tongue of land which cut the distance to the booming ground several kilometers. It was called the "cutoff" then, and still is. A "trip," or sheer boom, was strung across the river at the head of the cut-off, and incoming logs were turned into it. Piling was driven from the south end of the tongue of land to the east shore of the lake a short distance to create a man-made bay in which the "east" and "west" booms were constructed.

More pilings were driven into the man-made bay, and a look at it from the air would have revealed a labyrinth of pocket booms and pens surrounded by men walking on permanent "boom sticks," (long logs) shackled together in fours, with planking on top to serve as walks. And waving every which way were the long pike poles used by the men to push and pull the logs around to their proper pens.

As the logs floated into the sorting gaps, their ownership was established from the bark marks, but there were some logs which had no bark marks visible because the log had, in the course of the drive downstream, rolled over. The Oshkosh *City*

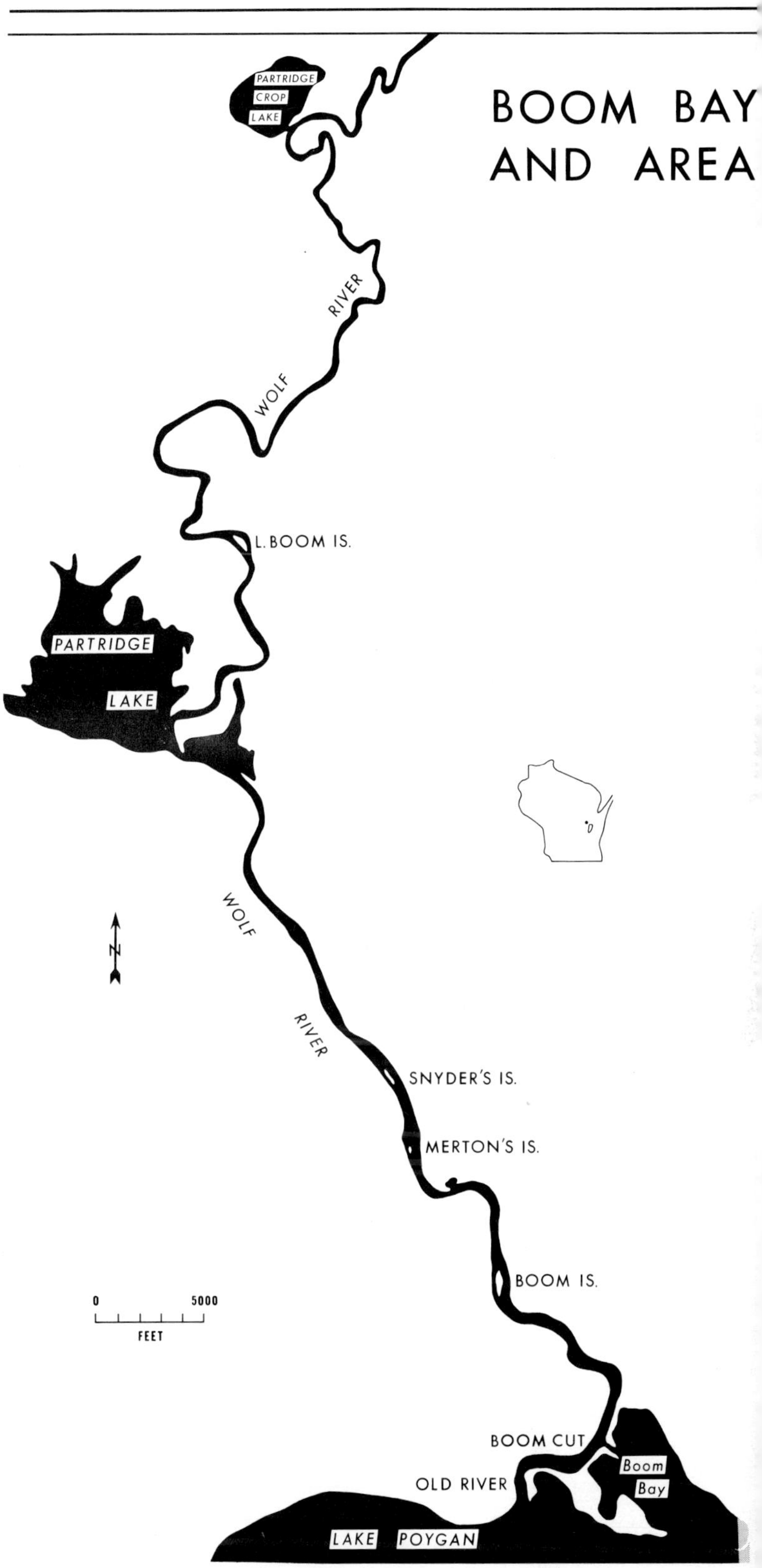

Map by Randall Rohe of lower Wolf River and approach to Lake Poygan where logs were sorted.

Times, in its July 21, 1868, edition, described the process this way:

> On arriving at the bay the logs, having been guided into one of the two booms on either shore of the bay, were subjected to the operations of catch markers. The logs did not always float so as to show the mark put on them at the landing, and to save time and trouble by the gap-tenders, catch marks were placed on the side of the log that floated up most. Meanwhile, the gap-tenders were stringing boomsticks along the race, a narrow passage formed by two double rows of boomsticks that reached from the head of the bay nearly to its mouth. At regular intervals a gap is left in the boom, and a gap-tender is stationed there to guide all logs marked in a particular manner from the race through the gap into the enclosure. Here the placers arrange them into rafts and they are towed to the mills.

The catch mark referred to here was actually a new, abbreviated bark mark, hurridly inscribed with an axe on the top of the log as it arrived in the first holding boom. The original bark mark had been inscribed at the landing before the log was sacked into the river farther up the Wolf and its tributaries.

The logs were also hammered at one end with a stamp hammer, as noted earlier, but this too might appear upside down by the time the log reached Bay Boom. To facilitate the passage of this log into the right sorting pen, there were men waiting at the head of the boom works ready to inscribe a "catch mark" if necessary.

These log markings were used in mortgage liens held against the owners or contractors of logs. The logger or jobber who had the contract to deliver a certain number of feet of timber often borrowed money to keep his operation going. He then had to provide the mortgager with a copy of his stamp hammer or log mark, and also what it was referred to in English.

For example, on the Wolf River in the 1870s, according to James E. Lundstad, there was an "I S I" mark which was called the "notch pothook notch." This was for a loan of $1,052 against one million feet of timber. Another lien was made out against the mark "I H I" or "notch h notch." Still another was a petition for a lien against three and a half million feet of timber which were coming down the West Branch of the Wolf River with the mark "II X M," called the "two notches cross M." Other marks included the "bucksaw over cross," and "double cross over crowfoot." Conlee Bros. Lumber Company of Oshkosh used two "q's" which, with two tails at the bottom, resembled a "w," and was referred to by the drivers as the "double-you-boot-jack" because it resembled a boot jack, a device found outside sleeping shanties to aid the men in pulling off their boots.

At Bay Boom, gap-tenders stood at the head of the boom to guide the incoming logs into the correct gap or sluice-way. The "placers" were crews that made up the rafts within the jam boom before the boom was opened to release the made-up raft for the journey across Lake Poygan and down the Fox into Lake Winnebago. The placers also assembled the logs into a "fleet." (The word "brail" was not used on the Wolf.) A fleet was made up of a string of eight to ten rafts.

Meanwhile, there were always stray logs without bark marks of any kind floating downstream. These were caught and shunted into a "bull-pen" at the sorting works, and if no one could prove ownership, they were sold at auction by the boom company.

Most of the mill owners in Oshkosh and Fond du Lac had their own tug boats, and when a fleet was ready, a tug ran the length of a cable out and the order "down grouser" was heard. The line was wound up on a drum powered by the tugboat steam engine. The "grouser" was a long, upright stick, four by four inches thick, with a steel point and a ratchet on one side for hoisting.

Mrs. Chester W. Smith, a pioneer of Winneconne, recalled the halcyon days at Bay Boom and the log rafts that went past her door. Writing in the *Wisconsin Magazine of History* 1972), she said: "*The Badger, S.W. Hollister, Ajax, M.D. Moore* and others were ahead, behind, up and down on each side of the rafts, whistling and responding to the call from the raftsmen, 'snub 'er,' meaning to lower the grouser and tighten the line connecting the raft with the tug. And those raftsmen with their long pike poles guiding the rafts jumping from one rolling log to another!"

The "grouser" mentioned here was situated in the middle of a tug boat and when released, the

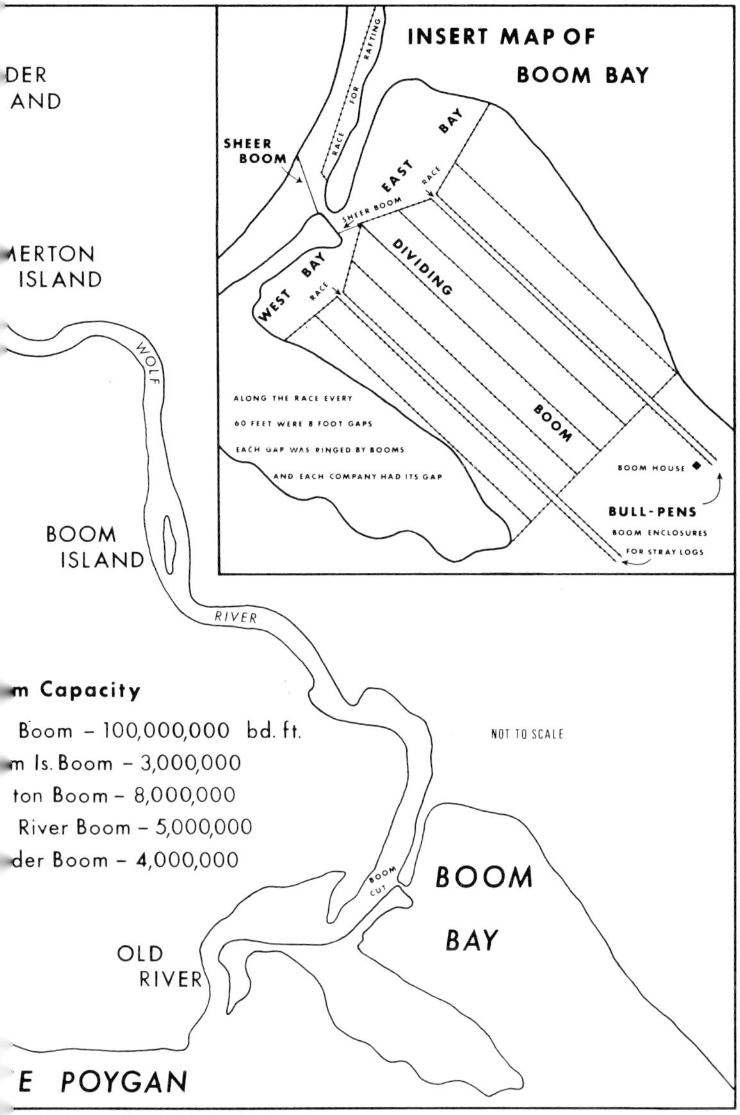

The second largest booming works in Wisconsin was located at "Boom Bay" (or as "Bay Boom"), at the mouth of the Wolf River on Lake Poygan in Winnebago County. Above, a diagram by Randall Rohe which explains some of the details of the works and rafting pens.

lower end dropped into the lake bottom. With the grouser firmly anchored in the mud, the tug boat reeled up a cable attached to the log raft, and in this manner slowly pulled the raft across Lake Poygan, one cable length at a time until reaching the current of the Wolf River at Winneconne. In a word, the tug boat did not pull the raft; the winch on the tug pulled a long cable attached to the raft.

In the peak of season, up to eight hundred men were employed at Bay Boom, either in the sorting works, or making up log rafts, or operating tugs. Company officials, clerks and time keepers lived in "Boom House," a one-story hotel resting on piling high enough to keep the floors dry. Workmen lived in nearby shanties and in the villages around the lake.

Thomas Wall was boom boss at Bay Boom for many years in the 1870s and 80s, and when he died, a big funeral was held in Oshkosh. Aside from the many floral wreaths sent by friends, there was one large floral log with the words "Rafted Out."

Today, it seems difficult to believe how many millions of board feet of timber were run down the rivers of the Wolf River Valley, even as early as 1864. There was by this time, of course, a tremendous demand for lumber products in a nation at war and a nation ready to explode in all directions of economic growth after the war. There were three drives on the Little Wolf River that year which alone amounted to nearly twenty-five million feet of timber, all of it probably cut in Waupaca County. (Logging farther up the Little Wolf in Portage and Marathon counties did not begin until the 1870s.)

F. James George, a lumberman, and grandfather of William Alft, entered in his notebook the total number of board feet that came down the several rivers into Bay Boom in 1864. When he speaks of the "Shawano drive" he probably refers to logs from camps as far north as the Indian Reservation, but not beyond, because logging north of the Reservation did not begin until after 1868. Here are George's figures for the Wolf and its tributaries in 1864:

Shioc, Northport & Fremont drive, bay scale [meaning the scale recorded at Bay Boom]	4,968,000
First Little Wolf drive	9,930,706
First Embarrass drive	6,765,000
Camel's and Dodge's drive	7,468,044
Geo. Lawrence drive at Snyder boom	3,001,355
Second Little Wolf drive	12,023,964
Shawano drive	22,435,000
Third Little Wolf drive	2,815,426
Red River drive	12,831,298
Second Embarrass drive	13,559,423

The Red River and the Embarrass drives could both be lumped with the Shawano drive, but

Tugs in Lake Poygan were used to tow the log rafts assembled at Bay Boom across the lake and down the Fox River into Lake Winnebago. But the tugs were not big enough to tow a raft across the lake unassisted. Instead, a technique was developed whereby a timber with a ratchet, called a "grouser," was plunged through a hole in the center of the bow into the lake bottom. A cable went back to the log raft and the cable was then hauled in by a steam winch on the tug. After the raft had approached the tug, the grouser was cranked up on the ratchet and the tug moved forward again to the end of the cable, and the order "down grouser" was heard again. In this picture, the grouser is the ratchet timber rising above the wheel house. It had a steel shod tip which extended to the bottom of the hull like a center-board, and was raised and lowered by machinery. When loosened, its own weight plunged it to the bottom with such force as to drive it securely into the lake bed. The first grouser was allegedly developed in the summer of 1848 by David Humes of Omro (Winnebago County). Webster's *International* dictionary defines it as a "pointed timber attached to a boat."

apparently the logs that came out of these two tributaries were owned by lumbermen not associated with the main Shawano drive.

Jack Hunter of Fond du Lac became a legendary figure on the Little Wolf River as a foreman for Moore & Galloway. He was a huge man and often boasted that he had forgotten the last time he had seen his shoelaces. When he visited Waupaca in 1880, *The Republican*, in its August 13th issue, had this to say of him:

> [Hunter] personally known to everybody who ever owned or handled a saw log on the Little Wolf, was in town last Friday. Moore & Galloway & Baker of Fond du Lac owned a large amount of pine in Town 26, Range 11 [Germania] but between it and the usual logging waters of the Little Wolf, was [sic] twelve miles of rapids, and Jack undertook to make it possible to run that pine out, and he did it with powder, teams, and derricks. He blew out that 'ere, and pulled out that 'ere, until he had a twenty-foot channel through the rocks the whole twelve miles, and got the logs to the boom this year in June. Jack is a whole team and a mule to spare.

The town of Germania lies in the southwest corner of Shawano County, and Hunter probably was clearing the rocks and windfalls from Comet Creek to reach the Little Wolf. Earlier he had cleared out the windfalls from the Little Wolf in northwestern Waupaca County. One who remembered hearing the blasting as a boy was the late Ole P. Dobbe (Portage County). Although he lived two miles from the river, he could hear the noise of the blasting through the forest. Later in life, he learned that it was Jack Hunter who made all the noise.

After twelve dams had been built on the Little Wolf River, it was found that there was still one

The pictures at left and below were taken on Lake Poygan, but both fail to give an accurate idea of the vast sorting works that was established at "Bay Boom" for sorting the thousands of logs that came down the Wolf River every spring. The photographer had no way of mounting a hill or bridge to get a birds-eye view of the works because the terrain around the bay is flat.

bad stretch of water between the Middle Dam (Portage County), and the Tim Crane Dam (Waupaca County). Hunter built a dam on the range line between the two counties and it is referred to in early texts as "Hunter's Improvement." A later generation, unaware of Hunter, called it the "Improvement Rapids." The embankments and part of the bulkheads were still visible to trout fishermen in the 1950s.

The late Hans Lee (Rosholt) remembered big Jack Hunter, for he had seen him while driving the Little Wolf in the 1880s. Lee believed that the last big pine drive came down from the Upper Dam in 1896. In the next several years some scattered lots of cedar and hardwood followed, but that was the end.

Recalling those exciting days, Lee told the author: "It was always easier to ride a log in quick water than in dead water, and a log runs faster in the river, you know, than the river itself. When you let the water out of the dam, first you have to let some of the "head" out before you start sluicing, so the head has a good start, but the logs always catch up to the head every time. I never saw it to fail."

It may be conceded, as Lee suggests, that the logs *in the middle of the river* caught up with the "head," but it would be impossible for the logs to outrun the *average flow* of the river.

CHIPPEWA RIVER
Holcombe to Chippewa Falls

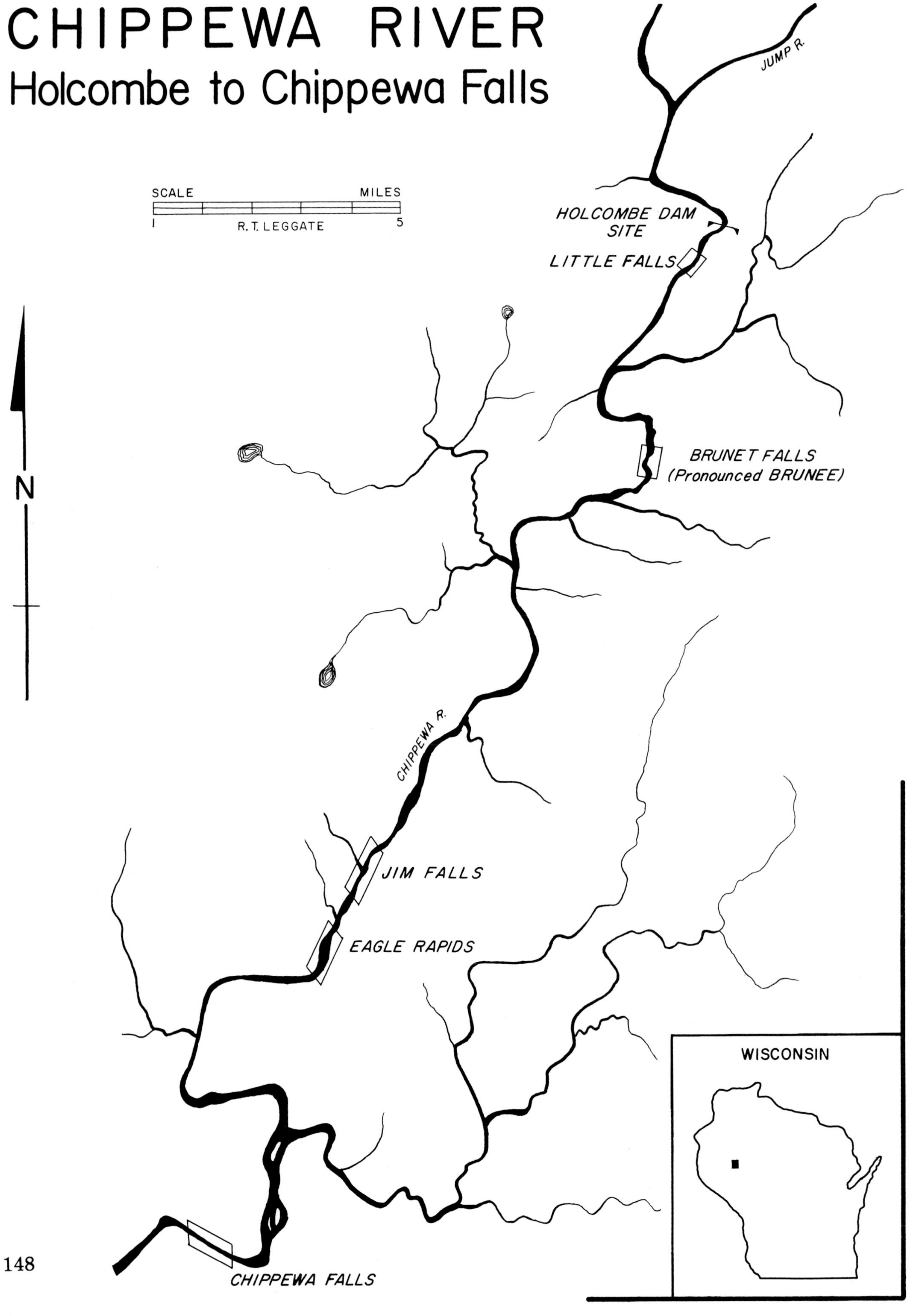

The "Chippeway" drive

On the Chippewa River, the technique of driving logs was not much different from that on the Wolf, but there was one interesting difference in terminology. J.L. Whitehouse called the crew on the Wolf that went ahead with the drive the "jam crew." But on the Chippewa, William Hoyer, last survivor of the Chippewa drive, said that he was a member of the "beat crew." Asked to explain this, he said:

> Well, there's always the crew they call the rear crew. When there's a rapids of dams or something, there were about ten or twelve men that went ahead, miles down the river, keeping an eye on things. They call them the beat crews. Best job to get in them little [beat] crews. Sometimes you don't do nothing. If the water runs good, okay, but if the logs pile up in the middle or something like that, you got to get out and straighten them out. The rear crews are cleaning up everything, working all day long. Here the poorest drivers were put to work because they couldn't take care of themselves. The boss wouldn't send anyone ahead for the beat crews who couldn't take care of himself. You started out in the morning with a tea bag, or lunch bag that you throwed over your shoulder, a tea kettle behind your suspenders and away you went. The tea bag was actually a tobacco bag but we filled it with tea leaves .... There was a fellow by the name of Jack Sugars, a great friend of mine. Big Jack, we called him. He was on the drive some damn place, I forget where, and Billy Gimble was the boss, and he sent Jack up to make the tea and lunch, and Sugars had a tobacco sack and a tea bag and made the mistake of throwing the tobacco in [the kettle]. And the men never knowd the difference! That's a God Almighty's fact. They drank it and never knowed the difference. Them guys was so full of pork and beans, you .... When you were on a beat you'd take two guys and you'd bring two logs together and you'd drive your pick (peavey) into the other log from the one you're sitting on, and the other guy on the other log would drive his pick into my log and we'd make a raft. I remember one time me and another guy was riding along nice and it came on twilight and all to once there was a tree leaning over, limbs right to the water. We were trying to get by that tree and we both got up on our log and the logs spread apart and the other guy went in and I jumped for the tree. The other guy had that tin can on the back of his britches for making tea, and that filled with water and he couldn't make any headway. But he managed to grab a branch of that tree and hold on till I could get ashore and find a branch to drag him in with. And we both lost our peavies there. Someone found mine later, I heard.

For the crew "bringing up the rear," there was no steam winch to lift a log off a sandbar or away from a wing. Eight to ten men, depending on the size of the log, waded into the water, hooked their peavies into the log and either lifted it clear or dragged it away from the bar into the water. Oxen or horses were also used to move logs from the banks farther back and in the bayous.

Logs on the Chippewa River were run all the way down from the Thornapple in Sawyer County to the Mississippi. The rapids at Chippewa Falls, after which a city is named, were dangerous, since there was a pitch here of about ten meters in less than four kilometers. A rapids upstream from Eau Claire was also a hazard in the early days but the rocks were submerged by construction of the King dam at Eau Claire.

The biggest threat to navigation on the Chippewa, however, was at Little Falls near the present village of Holcombe. Here a dam was built, the widest for log driving in Wisconsin, with no fewer than twenty-eight gates. And it was below this dam that a bateau of drivers was sucked into the fast-moving current and drowned in 1905. A big log jam had developed and crews were rushed up from Chippewa Falls to break the jam. Anxious to get to work, sixteen men hurried into the river in a bateau to attack the jam, some carrying bottles of whisky from the night before, and everyone in a holiday mood. As the impending tragedy approached, five men leaped to safety on the log jam, but the others remained in the bateau which overturned and threw the men against the rocks and logs like toys in a bathtub. And among them was "Whitewater Ole" (Horne), one of the most skilled drivers on the Chippewa River. Another victim was Louis Cokey who a year earlier escaped death at Jim Falls when his brother Mike drowned. Cokey was the fifth member of his family to drown in river accidents.

Whitewater Ole always boasted that despite the hazards of his profession, he would die in bed. His body was found floating not far from his house in Chippewa Falls, and two or three of the farmers who had taken employment as drivers and drowned in the same accident, were found near their farms along the river. A legend grew that all the bodies of the drowned men were picked up next to their farms on the river, as if they were going home.

Jim Falls and Eagle Rapids, both danger points on the Chippewa, lie halfway between Chippewa Falls and Holcombe. Another danger point was at Paint Creek Rapids, a short distance downstream from Eagle Rapids, but this is now submerged by the dam which created Lake Wissota.

William Hoyer believed that the last drive on the Chippewa took place about 1915. On the Flambeau the last logs were probably run in 1926. This was not a proper drive but was made to pick up strays or "dead heads" left behind from previous drives.

The last drive out of the Chippewa Flowage was probably run in 1917, but no one seems to remember where the logs were going.

The last drive on the Wisconsin River may have been run in 1916 when the John Week Lumber Company drove from a landing on the upper Rib River down to the company sawmill in Stevens Point.

William Natzke (Merrill) believes that the last drive on the Rib River was run from April to June in 1921, for the Wausau Paper Mill. The logs were mostly hemlock, but the crew built a wanigan to make the run "official." Wages at that time, according to Natzke, were six dollars a day or forty-two dollars a week, with no day off for Sunday.

Around the turn of the 20th century, Frank Dolan, a one-time "river pig" on the *Chippeway* drive, composed a bit of doggrel called "Footprints on the Sands of Time." It is worthy of inclusion in the narrative here because it gives the names of some of the most well-known early lumbermen and drivers on the Chippewa River. The poem was preserved in the family of the late Roy Martin (Winter) who inherited it from his uncle, Edward ("Eddy") Martin, also a river man whose life was spared on that tragic day at Little Falls in 1905. The time period seems to cover the 1870s and 1880s. Some of the men mentioned later became lumber tycoons and prominent businessmen at Chippewa Falls.

William Hoyer

Reminiscing with the author in 1976, Hoyer remembered Malcolm Josie who "lived just this side of the bridge in *Chippeway* Falls. Old Man Josie and Mrs. Josie used to sit on the porch when I come home from school . . . ."

Pete Lego had a livery line when Hoyer knew him: "He was a French-Canadian, part Indian." As to the Mandelerts, Hoyer could not recall their first names but he said they had a store in Chippewa Falls where the jacks bought woolen goods. "And I used a wool undersuit and trousers, no shirt, and when you fell in the river, the water drained out of the wool and you'd be dry in couple hours."

Sam Babb, Hoyer recalled, got an island on the Flambeau River (near Big Bear Lodge) named after him. Sam had a boy called the 'Chippeway Kid.' "

"J. Hokenbrock? He made that park [in Chippewa Falls]. Chippeway Lumber & Boom donated the land and William Irvine got his name on it. Hokenbrock had the horses that landscaped the park."

"Bill Fowlds," Hoyer continued, "lived in the addition by the graveyard on Stanley hill, east side of town. Came from Scotland. And Bill Dolan was a brother to old what's his name. They went to hell-an'-gone up there in Minnesota. And Jack

Footprints on the Sands of Time

FRANK DOLAN

In reverie I wander to the days long past and gone,
Like a miraage I can see the big sawmill,
And the river full of saw logs that the lumberjacks brought down,
When I think of them it gives my heart a thrill.

Oh, how the boys would celebrate when their winter's work was done,
With a suit of clothes C.T. Nelson made they ready for the fun,
The streets were lined with lumberjacks and everyone seemed glad
To shake the hand and welcome home the jovial shanty lad.

Most of the boys have passed away but we never will forget,
Jim Taylor, Dan Crowly and our late friend H. Vinette,
Then comes Malcolm Josie and the Caldwells, Dave and Jim,
With Jack Bell, the superman, we all remember him.

I remember well Pat Gaynor, Chick Stiles and Hugh McPhee,
Tom Gaynor, Ed Rutledge, John Morning and Mike McGee.
Big John McDonald, so muscular and so tall,
Mart Cummings, Mr. Thomas Crooks and then comes Bateau Paul.

I never forgot Pat Agnew, Slim McDonald and Pete Lego,
Denny Carrol and the Mandelerts, it seems so long ago,
George Gilmore and Dunc McCoy, whose name was good as gold,
Sam Babb, J. Hokenbrock and Hughie Cameron and good old Billy Fowlds.

Stickney and McPherson, Bill Dolan and his brogue,
Jack Whidden and Jerry Day and that pioneer, Dan Hoag;
And here's to Bill Mullen, that good old shanty chef,
And also big Tom Murphy and Flavian LeBoeuf.

Of course there were many others with hearts as true as steel,
Bill Irvine, August Mason and our good friend Gene O'Neil,
Every one a sturdy pioneer, so jovial and so kind,
And they called themselves the shanty boys while falling down the pine.

These good old logging days are past, and we'll never see them more,
The last loggers I remember were Charlie Stinson and Art Gore,
Oh, no, there was Smith and Hanson; now I'll have to make this rhyme
With Erick Myrman following footprints on the sands of time.

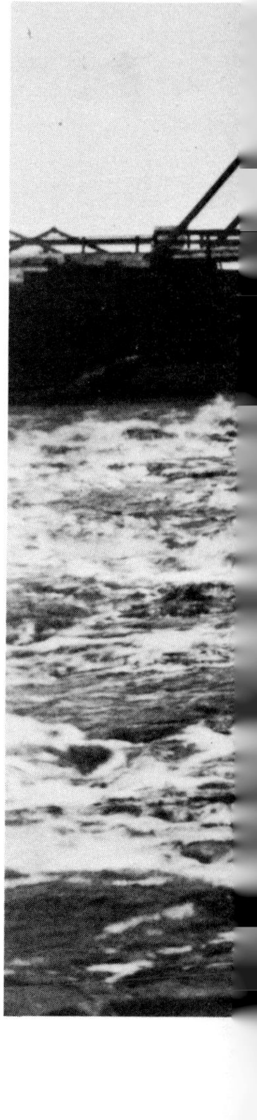

The Nevers Dam on St. Croix River as seen from Wisconsin side, September 13, 1897, in photo by S.C. Sargent of Taylors Falls. Dam created flowage ten miles upstream and was used for sluicing logs destined for down river sawmills at Stillwater, Hudson and Mississippi River towns. Last log drive went through dam in spring of 1913. In 1914 it was acquired by Northern States Power Company and replaced with concrete dam 740 feet wide. Photo at lower left, opposite page, shows concrete dam washout on Wisconsin side in 1954, when dam was abandoned and demolished. Dam seen here was built in 1889-90 by St. Croix Dam & Boom Company, 634 feet wide. It had twelve waste gates, all Tainter construction, twenty-four feet wide and sixteen feet deep, in addition to three submerged "bear trap" gates, one eighty feet wide, one twenty-four and one twenty feet, making this probably largest "bear trap" construction in North America. (For illustration of bear trap see *Lumbermen on the Chippewa.*) Bear trap gates were located where steel bridge stands in pictures here, but they had already been abandoned when picture was taken in 1897.

Whidden, they called him 'the bull of the woods,' a well known character. I ran beat with him on the lower Chief [River] after he tamed down. But that Jack Bell, they had a saying about him, 'fourteen men on one side of the log and Jack Bell on the other.' "

As for Bill Millen, "they called him an old shanty chef, but he didn't know nothin' about chefs. He was probably a bull cook. Kept the fires going."

"Tom Murphy? I knew him well. I worked for him up there around Winters (Most lumberjacks referred to this village in the plural, not as Winter (Sawyer County.) Tom was a boss. His son was Paddy Murphy."

"Flavian LeBeouf was one of these French Lumber Company guys in Chippewa Falls. August Mason was the father of Johnny, Alex, and Burt, and I went to school with those boys, and when I began working in the woods, I worked five years for Mason who was logging for the Rice Lake Lumber Company. He had a first-class outfit. Usually a hundred men to a camp."

Gene O'Neil, Hoyer continued, was "the fellow they called Red Horse O'Neil because that's all they had to eat in camp. We had a nickname for everything. There was the "Gut & Liver line [railroad] up there towards Bemidji. Red horse, that's not just a fish. It's the same as bologna. Some kind of meat they served, I'm not sure, probably horse meat. Seems to me we worked near Bemidji for the White Pine Lumber Company run by Moses and Gaynor."

(William Moses and Pat Gaynor, mentioned in the poem, were a partnership for several years, and before moving to Minnesota, one of their big jobs in Wisconsin took them to the Thornapple River in the winter of 1907-08 where they signed a contract with Chippewa Lumber & Boom to move all the logs stranded on the Thornapple River, after Cameron Dam washed out, six miles east to Babbs Landing on the Flambeau. A steam hauler was used in this operation.)

Hoyer failed to recall Hughie Cameron, who is mentioned in the poem. Cameron owned the property on which Cameron Dam was located on Thornapple River. After Cameron died, his widow, Jennie Cameron, sold a quarter section of this tract to Hattie E. Dietz, wife of John F. Dietz, and trouble began when Dietz learned that half of the dam lay on his quarter section, the other half on land owned by Chippewa Lumber & Boom.

Hoyer worked for the firm of Moses & Gaynor in Minnesota on the "Gut & Liver Line" where "Red Horse" O'Neil was walking boss. "And I quit," he said, "and was comin' down the railroad and I meets this Mike Murphy. He was woods boss for Moses, and the landing was all blocked up. I knew Mike from Hayward, and he says why don't you stay and work for me? Okay. He says the landing's all blocked. So I worked a couple of months for Moses. I think it was called the White Pine Lumber Company, before the First World War. After the war I run across this Moses man out in Dakota. I was standing on the corner in a little town called Herman where I'd been out working for a farmer, threshing or some damn thing, and here comes Moses in a rig. He wanted me to work for him again and I said, 'Oh no.' He was kind of a hypocrite. Before when I was working for him, he'd preach every Sunday to the crew. Had a regular service, and Mike Murphy was there, the boss for Moses, and I was sittin' alongside Mike and he said, 'Just listen to him! I'll never forget that service. Him preachin' to a bunch of guys never been in church before."

The Chick Styles mentioned in the poem fell down a stairs in his home at Chippewa Falls and died of injuries. Hugh McPhee moved out to British Columbia. When Hoyer knew Charlie Stinson and Art Gore, they were logging on the Clam River, but in the early years of the 20th century, Stinson was a foreman for the Kaiser Lumber Company. A spur, called the "Stinson Spur," was built south from Winter to the Thornapple River where Kaiser logged several winters.

Edward Rutledge, who came to the Chippewa Valley, via Canada, in the 1860s, was born in Ireland. Hoyer referred to him as "one of them big shots with Chippewa Lumber & Boom." This was a wholly-owned subsidiary of the Mississippi River Logging Company of Frederick Weyerhaeuser.

By 1887, Rutledge had risen to the top of the heap among lumbermen and had acquired the

The Soo Line railroad bridge was built almost directly over the old logging dam at Chippewa Falls. Under the bridge are piers to which the "boom sticks" were secured, although the Soo Line bridge stood on its own footings and did not rest on the dam. At the right is a chute which seems to be made of heavy planking and was used to run the wanigans over the dam. The logs came over the slide in the center where the water rushes through.

A caption on this picture reads: "Old Paint Creek Dam," but most of the dam had already vanished when the picture was taken. Earlier, Paint Creek joined the Chippewa River a short distance upstream from Chippewa Falls. The view shown here is now obliterated by the new dam which created Lake Wissota. (L.G. Arnold, a government engineer from Minnesota, named the lake.)

residence of James Monroe Bingham, later a lieutenant governor of Wisconsin. The house and outbuildings were extensively remodelled. Rutledge died in 1911 at the age of seventy-eight, and four years later, Dayton E. Cook, a local attorney and later judge, acquired the property from the Rutledge estate. In 1973, a granddaughter, Mabel L. Cook, deeded the property to the Chippewa County Historical Society, and it is now a city museum called the Cook-Rutledge Mansion. It is considered a fine example of High Victorian-Italianate architecture and is listed in the National Registry of Historical Places.

Rutledge was preceded in death by his wife, Hannah, and in her memory he built the Hannah M. Rutledge Home for the Aged, and also donated and endowed the Edward Rutledge Charities.

Five bateaux and crews prepare to attack a "center" (of logs) and free them on the Chippewa River, early 1900s. A.A. Bish, a studio photographer of Chippewa Falls, took the picture below the one-time iron bridge which spanned the river about two kilometers south of the dam at Little Falls (Holcombe). William Hoyer, born in Chippewa Falls, was the only survivor of this entire crew before his death in 1979, and he stands in the bow of the bateau at the rear, his hands on a pole. He spent most of his adult life in business at Cudahy.

The crews shown in the picture were employed by Chippewa Lumber & Boom Company. The company's most trusted river boss in the 1890s and probably down to the end of the log driving era on the Chippewa in 1911, was Archie Smith who may be seen in the bateau at the right, standing, in dark suit with stick over his right shoulder. In 1905 Smith was standing on the dam at Little Falls, watching operations below where a jam had formed. When a bateau with sixteen men left shore to attack the jam, he looked on, helplessly, as the bateau overturned and eleven men were drowned, some of whom are probably in this picture.

The Chippewa River makes a great horseshoe bend north of Eau Claire. A dam was built in the city to create "Dells Pond" below the bend for booming and sorting logs. These two views (left) of the same works, taken from opposite directions, offer a clue to the great harvest of logs that was sorted here every year for more than forty years.

Below: two-gate dam on a lake probably in northern Wisconsin around the turn of the 20th century. The two men are not in working clothes and this may be a Sunday, but one of them might have been the watchman who opened the gates when a spring freshet came on, or closed the gates to build up a "head" of water when a log drive was under way.

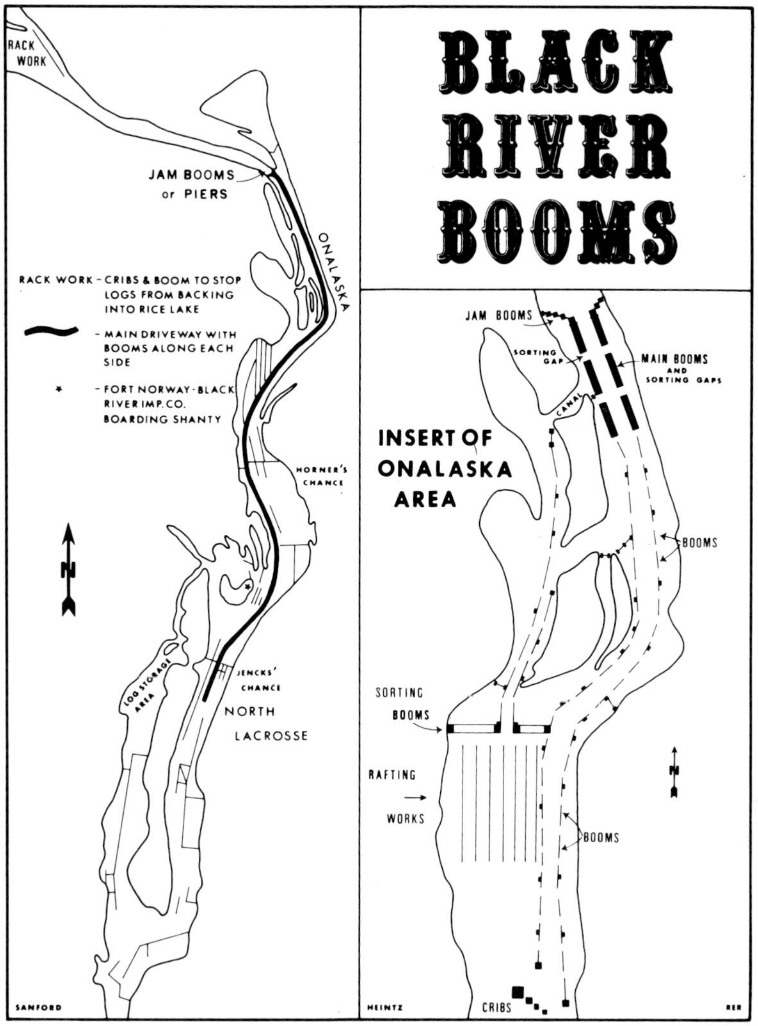

River drivers stop for lunch, probably on Black River in Jackson County. This is a landing and the logs in the foreground will probably be sacked into the river from this point. Note the "gig" path running up the river at the left, used by the drivers "gigging" back and forth between the drives.

A map by Randall Rohe shows sorting works on Black River in north La Crosse.

The once famed Cameron Dam on Thornapple River (no relation to the village of Cameron) in a picture taken before 1906, the year one wing of the dam washed out in spring flood because crews of Chippewa Lumber & Boom Company, who were camped nearby, were unable to tend the gates when threatened by John Dietz, the so-called "Defender of Cameron Dam." This was a three-gate dam, built in the late 1870s, but in 1904, after the trouble with Dietz began, it never sluiced another log.

A small bateau fights its way down Namekagon River near Hayward. In 1912, the Marshall-Wells Hardware catalog advertised a bateau thirty-two feet long, for three stroke paddles, weighing three hundred and fifty pounds, for $50. A four-stroker, thirty-six feet long, cost $70. The seven foot paddles were made of hard maple and weighed four and three quarters pounds.

The stage of water is running high, and these men are sacking logs left behind by the main drive. The man on the left is Mike Barney. The photo was probably taken to Potato River near Upson (Iron County). In the foreground lies a boom stick which has been used for a boom on the flowage created by the dam.

A boom stick preserved at the Menominee Logging Camp at Keshena.

The photographs, above and below, were taken on Red Cedar River north of Menominee for Knapp, Stout & Company. The crews in the four bateaux (above) prepare to attack a "center" (of logs) seen in the lower picture.

Above, a crew for Knapp, Stout & Company sacking a log on the **Red Cedar River**, and below, a wanigan on the Big Eau Pleine, probably for the R. Connor Company of Stratford. Here the men have gone ashore for the night and slept in the tent on the embankment. It was the duty of the cook and cookee to erect this tent wherever the drive stopped for the night. A table made of planks has been erected in front of the tent where the food is served, but there are no chairs.

The pictures at left and below were taken on Rib River near a community called Interwald (Taylor County), probably in 1903. In the picture at the left, the men are sacking the logs from the rollway, that is, pushing them into the water for the drive down river. They are using both pike poles and peavies. In the picture below, two of the crew attempt to break up a minor jam. It is early spring and the snow still lies on the ground.

Right: The plaque below advises the viewer that this is a driving crew on the Wisconsin and Rib rivers for "Week's" (the John Week Lumber Company of Stevens Point) in 1903. All but one are holding peavies, and the man in center with pipe is no doubt the straw boss.

A wanigan on Rib River at Interwald. The small bateau at right probably trailed the wanigan. The cook, seated, and the cookee at his side, both appear to be wearing white shirts with ties. Man next to the cookee holds coffee can for benefit of photographer.

Two Indians, probably Chippewas, on Rib River drive. The one at left wears special buttons on shirt and sleeve.

Beef slough

For nearly two decades, from the early 1870s into the early 1890s, the largest booming works in Wisconsin, if not in the United States, was located at Beef Slough lying to the north of Alma along the east bank of the Mississippi River and extending northeast along the south bank of the Chippewa River, at least twenty-five kilometers in extent. No one can be sure how it got its name, although bunk house historians are ready to supply the answer if asked to do so. Something about buffalo meat that got left behind in the slough, etc. etc.

When the logs in the Chippewa River approached the slough, they were diverted into it by a sheer boom and from this diversion point, they floated lazily down the slough into the Mississippi where they were diverted into another slough along the east bank for sorting and making up into brails.

The company that built the Beef Slough works was incorporated in 1868 as the "Beef Slough Manufacturing, Booming, Log Driving & Transportation Company," usually abbreviated to the "Beef Slough Company."

The year after the boom company was organized it handled nearly six million board feet of timber; by 1875, under new management, the figure jumped to 129,066,630 feet, and a decade later in 1885 to more than half a billion.

In the beginning, the Beef Slough Company was having difficulty getting its logs down the Chippewa River because of delaying tactics created by lumbermen at Eau Claire and Chippewa Falls. The company, which had invested heavily in building the Beef Slough works, found it difficult to maintain a "cash flow" sufficient to meet expenses, much of the problem lying with the mill men on the Chippewa. In 1870 the stockholders leased the Beef Slough Company to the Mississippi River Logging Company of Frederick Weyerhaeuser.

Before Weyerhaeuser came to the rescue, the Beef Slough Company had asked the state legislature for a special charter to deal with the mill owners in Eau Claire and Chippewa Falls, but the bill was defeated. This was considered a victory for the mill men and they proceeded to close their booms in Eau Claire. But a last-minute rider to another bill, totally unrelated to logging, passed the legislature and reinstated the privileges sought by the Beef Slough Company. The latter then sent a force of more than one hundred men to Eau Claire to bring down the logs that were being held up there. To do this they had to break the boom at the Dells, just north of Eau Claire, on the Chippewa River. The number of logs that were released is unknown, but the county sheriff and a posse arrived on the scene before too long and arrested the ringleader and a number of other drivers who were unable to get away. This was an incident in the so-called "Beef Slough War," although no one had any weapons other than jam pikes and hand spikes. But "prisoners" were taken, and, perhaps that is why it is referred to as a "war."

An uneasy truce was reached between the contesting parties of lumbermen, but this and other difficulties forced the Beef Slough people to turn to Weyerhaeuser for more capital and, most important, better management. Finally, in 1880 a new arrangement was worked out called a "pool," whereby the Mississippi River and Chippewa River mill men were to share the logs coming down the Chippewa River on a percentage basis, but contemporary accounts fail to explain how this sharing was managed.

Since there were different mill owners on the Mississippi at Winona, Clinton, Rock Island, Dubuque, and other cities, the logs had to be made into rafts for running downriver to the mills. The first method used was to make them into rafts with steering oars three meters long, fore and aft. In *Harpers* for October 1855, T.B. Thorpe has an article called "Remembrances of the Mississippi," mostly about life on the lower Mississippi, but on page thirty-nine appears a small engraving of a Mississippi raft. The accompanying illustration has been twice enlarged for this book. However, there is little in the engraving to suggest how the raft was assembled.

The log rafts at this time were small, and they were difficult to assemble for the amount of time expended. A new system was then developed by forming logs into "brails," allegedly named for a

An official marker erected by the State Historical Society of Wisconsin which stands a short distance north of Alma on the Mississippi River, calls attention to the great logging epoch that transpired here when Beef Slough was the largest sorting works in Wisconsin, if not in the world.

foreman named Brayler, who first thought of the new system of binding the logs together by using wooden binder poles. "Boom chains" and "boom plugs" were made by the company blacksmith to further secure the brails. Finally, a chain of logs, or "boom sticks," linked together on the outside, went clear around the brail to act as a further tie-in. The logs circling the boom were held together with iron links and were prevented from spreading by galvanized wire. When everything was secured, a brail was at least five hundred feet long and forty-five feet wide. Six brails coupled together made a "Mississippi raft."

A document in the Laird Norton Company

This was apparently the heart of the Beef Slough sorting works on the Mississippi River north of Alma. It may be that this crew is fitting logs into a compact mass that will be made into a "brail," that is, an enlarged raft held in place by boom sticks on the outside, and prevented from spreading by galvanized wire lines spaced at intervals across the deck. A brail was five hundred and fifty feet long and forty-five feet wide, and six brails, coupled side by side, constituted a full "Mississippi raft." When a brail was completed, it was nudged into the open water of the river and guided by a steamboat astern, and a bowboat, tied crossways to the raft, at the bow. Neither the steamer nor the bowboat was expected to push or tow the raft. The river did that, but the raft had to be guided around the bends and sandbars. Oarsmen on the brail also helped swing the raft. In the picture above, the men in white shirts are the scalers with their rule sticks measuring the number of feet in the brail. Charges were then made against the owners for booming of the logs, that is, sorting them, storing them, and making them into brails. This picture and the three to follow were taken, circa 1890, by Gerhard Gesell, a pioneer studio photographer at Alma.

papers at the Minnesota Historical Society relating to the Ann River Logging Company, dated Stillwater, MN, August 22, 1889, reveals that after a brail had been scaled, it was given a number, and a report of the number of feet scaled in it was then sent to the Board of Trade, presumably in St. Paul.

But the brails were not guided downriver by long oars. A steamboat in the stern and a "bowboat" in the bow, guided them down the current and around the bends. Only a small crew was needed to accompany the raft, mainly to round up stray logs that might break loose.

At Beef Slough, where the brails were assembled, about six hundred men were employed. To accommodate these men, several sleeping and eating shanties were built nearby: Camp 1, Camp 2, Camp 3, Camp 3½, Camp 4, Camp 5, and Camp 6. Camp 1, which was located on Rieck's Lake, became the center of the Beef Slough lodging complex. It grew into a village with its own store, post office, company office, and later a depot on the Winona, Alma & Northern line.

A small crew was also camped at the "jam boom" and another at the "sheer boom," both located above the main Beef Slough works at the mouth of the Chippewa River where the logs were held pending movement into the sorting gaps of the main works.

The brails were made up within the Beef Slough booms and when completed, a swinging boom was opened and the rafts were allowed to slip into the main stream of the river where they were taken in hand by the steamboats and bowboats.

Before a brail of logs could be sent down to a sawmill, it had to be scaled, and the men responsible for this were the scalers who walked around the booms carrying rule sticks which had a small hook at one end. This hook was just big enough to catch the bottom of a log under water. Studying the figures on his rule, the scaler knew how much the log measured in board feet, marked it down on a sheet of ruled paper which at the top had a column where the "bark mark" of the various lumber companies was entered. This sheet was brought every night to the main office, where the clerks added up the figures and sent them on to the owners of the logs with the charge which was being made against them for sorting and storage.

One of the last men to remember the scalers with their rule sticks, the boom tenders with their calked boots, and the steamers blowing black smoke into the air as they stood off the banks waiting to take over a brail, was the late Theodore Buehler of Alma. Everything was going well for the people of Alma and Beef Slough, Buehler once recalled, until shortly after 1890 when Wisconsin passed a law to levy a tax on all logs processed at Beef Slough. Minnesota, on the opposite bank of the Mississippi, did not enact a similar law, so Weyerhaeuser pulled his Mississippi River Logging Company out and joined his fortunes with the Minnesotans. He built a sheer boom at the mouth of the Chippewa River to direct his logs away from Beef Slough and across the river to West Newton, where a new sorting works was constructed. However, it was only a matter of a few years before the great harvest of white pine in the Chippewa Valley was over, and then the West Newton works also closed.

The writer in *Harper's* (1855), mentioned earlier, describes the log rafts he saw on the Mississippi River as "crude masses of cypress timber, which find ready sale at the numerous sawmills in the vicinity of New Orleans." This suggests a raft made up on the lower Mississippi, since no cypress timber ever floated out of Wisconsin or Minnesota. "By an accepted law of the river," he continues, "everything is obliged to get out of the way of a raft. We don't know of any persons more independent than the first officers of these primitive flotillas. Their chief unhappiness is occasioned by the sneering remarks made by spectators relative to the speed of the rafts, and allusions to their propensity to leak and of the necessity of having the bottom pumped dry . . . ."

These remarks from bystanders along shore greatly inflamed the raftsmen who were, he says, "in a passion [to] hurl back abuse" at their tormentors.

Watching life along the Mississippi, the *Harper's* writer met a man on the river who told him, "I'd have nothing, bless God, to complain of, so long as hog meat is plentiful, and whisky keeps at a price whar a poor man has a chance."

Another part of the works at Beef Slough. This man holds a pike pole to push the logs into the sorting gaps and booms. He wears ordinary shoes, not calked boots, because he spends most of his time walking back and forth on the platform supported by piling in the river.

The picture above offer different views of the works at Beef Slough. Without aerial photography, it would have been impractical to photograph the entire works because it ran all the way north from Alma to the mouth of the Chippewa River, and up the left (east) bank of the Chippewa for some distance.

Left, a raft made up of several brails, moves down the Mississippi River, guided by a paddle steamer astern, and a bowboat, not visible in picture, in front of the raft.

In this picture of a Mississippi raft, both the steamboat and bowboat are visible. The bowboat (left) is in the bow, and is tied alongside. It will help the steamboat guide the raft by running its engines ahead or astern according to a signal from the steamboat. The bowboats also operated on steam and this one has a paddle wheel aft.

The *Helen Mar*, a packet for Knapp, Stout & Company of Menomonie, lies alongside at Dubuque on the Mississippi River. Packets were for company officials to sail on, for mail and limited cargo.

A crew is putting the finishing touches on a log brail, here seen securely bound by crossed wires and cables. In the stern rides the steamboat. Most of the men hold peavies and pike poles, but the man fifth from the right holds a long auger which is used for boring holes through the logs where wooden couplings are needed.

A sheer boom, or boom used to divert logs into gaps where "gap tenders" were waiting to push them into floating pens or holding booms, swings across the east channel of the Mississippi River north of Alma, seen in the distance. The sheer boom could be swung across a slough or river by manipulating the fins seen at the right of it, a technique developed by Levi Pond and his brother-in-law, James Allen, both of Eau Claire, in 1861. The river current did the work, not the men. This may be the only photograph available of a sheer boom with fins—a rare sight.

On Lake Superior

Right, a jack ladder and crane loading pulp in cars at Ashland for shipment to Wisconsin Rapids for Consolidated Papers Inc.

From 1923 to 1973—half a century—Consolidated Papers Inc. of Wisconsin Rapids moved two million cords of pulp wood into Ashland from Nipigon, Ontario, and Grand Marais, Minnesota. The logs were towed across Lake Superior in "pocket booms." In the picture above, a close-up of the boom sticks, and the beacon that rode astern to give the pilot in the tugboat a fix on the stern of the raft. Right, a close-up of shackles and link dogs holding the boom sticks together. The boom sticks used on Lake Superior were Sitka spruce from the Pacific Northwest, a species with more buoyancy than Lake State spruce.

A tug swings a "pocket boom" of pulp arriving from Canada into position at Ashland. The movement of the waves will actually bring the boom into shore. This particular pocket boom was under tow by *John Roen III*.

The boom companies

Although the Northwest Ordinance declared that all navigable rivers and waters were in the public domain, a hindrance to the free flow of logs in a river came with the organization of "boom" or "improvement" companies located in cities along the main river systems of Wisconsin.

A boom is best described as a floating pen of logs chained together to hold back or encircle other logs in a river or pond. They were called "booms," perhaps because when logs are strung together in this manner they are bumping and thumping each other from the motion of the current and this causes a hollow, bass-like sound. (On a log drive, too, logs were constantly bumping into each other. The "river pigs" in Maine remarked about the boom-like quality of this sound, and they said they could hear the logs approaching before they saw them, particularly at night after the wind died down.)

Since the waterways of Wisconsin were "freeways," anyone could use them to drive a log down to a mill, or, as it happened on the Black River and Chippewa River, all the way into the Mississippi River. To keep the logs from becoming lost, called "dead heads" or "prizes," that is, the property of no one, owners branded them on one end with a "marking hammer" or "stamp hammer." The stamping was usually done at the place of origin, often by the swamper, or scaler before the logs were sacked into the river. The handle of the stamp hammer was usually a little shorter than an axe handle.

At the height of the logging era in Wisconsin in the mid-1890s, there were several hundred log marks or "end stamps," registered with state inspectors in the several forestry districts. (A collection of stamp hammers, probably the largest in the United States, is on exhibit at the Menominee Logging Camp at Keshena.) But stamping a log on one end did not always guarantee ownership. River pirates stole logs stranded in sloughs, sawed off the ends and resold them with their own brand stamped on them.

In addition to the stamp mark on the end of the log, another mark was made in the bark with an axe on that part of the log riding highest out of the water. Since it was not easy to make curved lines

Photograph taken from railroad bridge, circa 1885, where t Black River flows into the Mississippi in north La Crosse. Sorti work booms, and crib piers extend for some distance northwar and most of the sawmills that were based in La Crosse can be se in the distance along the east bank of the Black River. The fi on the right is the D.B. Holway mill; next the John Packet Mi

with an axe head, the bark was chipped off in one spot to expose the wood and the marking was done, usually, in variations of straight lines. These were called "bark marks," or "water marks." Sometimes the bark mark and the end stamp were similar or identical.

Taking their cue from Maine, then, where the first boom company was allegedly organized in 1825, Wisconsin lumbermen organized similar companies to protect their logs and establish ownership of logs coming down the rivers. The first charter for this purpose was granted the St. Croix Boom Company in 1851, and the following year, the legislature approved the "Little Bull Falls Improvement & Steam Navigation Company" (Mosinee). Between that time and 1873, the state legislature granted franchises to about seventy-five boom companies in the state.

The men who worked on the booms to catch and sort the logs were counted in the thousands.

Io. 1 (two smoke stacks) and a machine shop (one stack); next, 'acket Mill No. 2 (three stacks); next the La Crosse Lumber Company's "Big Mill," and at the extreme left, the McDonald Bros. mill. Not seen farther left were Sawyer & Austin, and Crosy Iardwood Lumber Company. South of the Holway mill, but not isible (i.e. on the right), stood the A.S. Trow mill.

In 1856 lumbermen in the Chippewa Valley got a franchise to "dam all bayous and sloughs that make out of the Chippewa River, and to confine said river to its proper channel." In 1866, the state legislature passed an amendment to another bill which authorized the same lumbermen, now known as the "Chippewa River Rafting Boom Company," to collect a toll or "boomage" of twenty cents per thousand feet, board measure, for all logs, spars and square timber stored in their booms. (Spars were sold to shipping interests for masts, and square timber refers to "cants," that is, a log after the slabs have been removed on four sides to make a square. Many of these were shipped to England to be sawed into lumber in order to avoid a heavy duty on imported lumber.)

In 1863, Pembroke B. Stearns and others were authorized to erect a boom across the Wisconsin River opposite the present city of Nekoosa.

The Black River Improvement Company was authorized in 1864, and the act was amended in 1866 to allow the company to improve the river (Black) throughout its length and also to build dams and charge tolls for logs sluiced through the company dams.

An act to provide for the improvement of the Yellow River for log-driving purposes was passed by the legislature in 1873, but this probably refers to the Yellow River on the Chippewa, not to the Yellow River in Wood-Juneau counties. The act authorized any person or company to improve the Yellow River by clearing and straightening its channel, closing off sloughs, erecting slides, rolling and flooding dams, and keeping them in repair at all times for log-driving purposes. The owners were also permitted to collect a toll at "prescribed rates" only after $10,000 had been expended on river improvement.

Boom companies were also required to build a slide in the middle of their dams which could be used by anyone who wished to run a log or raft over it. On the Wisconsin River this slide had to be sixty feet wide. The slides, made of heavy planking, and anchored to cribs filled with rocks, acted like a chute on the river, and could be extremely hazardous unless they were built right and properly maintained. They were used mainly by the lumber raftsmen, not log drivers. There was a slide at Stevens Point in the 1860s, which was dangerous, but no slide on the entire river claimed as many lives as the one at Kilbourn (Wisconsin Dells).

The first booming company on the Wisconsin River was organized at Wausau in 1854 by Walter McIndoe and Company, who were authorized to maintain "certain dams, booms, and piers," provided that said dams, booms and piers did not obstruct navigation.

Abraham Brawley, after whom Brawley Street in Stevens Point is named, was authorized in 1859 to maintain a boom across the Wisconsin River, but there is no evidence that he ever completed it. On November 1, 1872, a group of Stevens Point business men filed notice of organization of the "Stevens Point Boom Company." The boom was to be located above Shaurette Rapids (just below the present Clark Street Bridge).

Famous log jams

There were several log jams in the history of the river drives in Wisconsin which attracted state-wide attention. One of the first occurred on the St. Croix River near Taylors Falls, about 1869. The Whitney Gallery of St. Paul managed to get a picture of it, probably the earliest log jam picture in the Lake States. Another bad jam occurred about the same time on the Chippewa River which backed up the logs from Chippewa Falls almost to Eagle Rapids.

In June 1886, there was a jam at Angle Rock on the St. Croix River that took several weeks to break up. Two Mississippi River steamboats, the *Arcola* and *A.B. Trow*, were brought in to assist, and sightseers came from St. Paul by horse and buggy to watch the operations, in which two hundred men and a hundred horses took part.

On the Jump River, a tributary of the Chippewa, a jam developed in 1879 which extended twenty-five kilometers upstream. In 1880 there was a jam at Wausau that backed up several kilometers. And that same year a big jam occurred at Overton's Rapids at the head of the Lily Dam flowage on the Wolf. Thirty million feet of timber were at stake, and it took a crew twenty-two days to break it up.

In 1888 there was a jam on the Wolf at Fremont which one writer described as "10,000,000 [i.e. board feet] of logs scattered all over the marshes and lake [Partridge] and their recovery was attended with great trouble and expense."

But smaller jams were commonplace. From a daily journal kept at the Little Falls dam (Holcombe), an anonymous writer explains that on June 20, 1885, a crew went up the Jump River to break a jam, but had no luck owing to low water. On May 26, 1886, a crew broke up a jam on Brunet River, and on June 3, there was a jam at Brunet Falls. In May, 1888, there was a jam above Little Falls dam, apparently in the boom works of the pond, that got worse and worse as more logs kept coming in. On May 21, in addition to a local crew, fifty-eight men were brought up from Chippewa Falls to help break it up.

One of the most famous log jams of the 19th century occurred on the St. Croix River in 1886, and took several weeks to break up. Salvage operations attracted visitors in horse and buggy as far away as St. Paul. Two paddle steamers with derricks were brought in to assist the men and horses. In the distance stands an early interstate bridge between St. Croix Falls, WI and Taylors Falls, MN. Below, a log jam in the same place on the St. Croix which took place circa 1889, but did not take as long to break up as the one in 1886.

This picture of a log jam on the St. Croix River, probably taken by a cameraman from the Whitney Gallery in St. Paul, circa 1869, is one of the oldest pictures of a jam in the Lake States to survive the 19th century. The condition of the logs suggests that they have been lying here for more than a year, or at least long enough for insect life and the sun to peel the bark. In the lower left stand two women in hoop skirts, and a man in dark suit and white shirt who faces away from the camera.

In draw-down of Wisconsin River at Wausau in 1984 these stone piers became exposed. Wood cribbing once held stones but cribbing has disappeared, leaving most of stone fill intact. These piers were used to anchor log boom that ran from base (bottom) of picture to top, and diverted incoming logs into left channel down to local sawmills and away from the right channel which leads to "Big Bull Falls," now a favorite for kayak enthusiasts. Photo shot from Bridge Street bridge by the author.

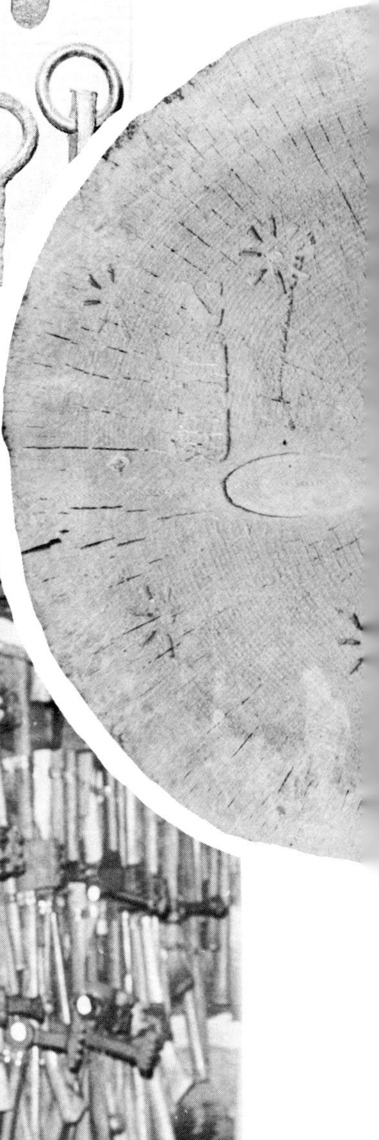

Above, raft shackles and link dogs used in connecting boom sticks on rafts or pond booms. Below, probably the largest collection of stamp hammers in the United States at Menominee Logging Camp at Keshena. Right center, a piece of cedar sawed from the end of a log found at the bottom of the Van Order Dam on Plover River (Portage County). The rosette pattern is augmented by lettering but too faint to be identified.

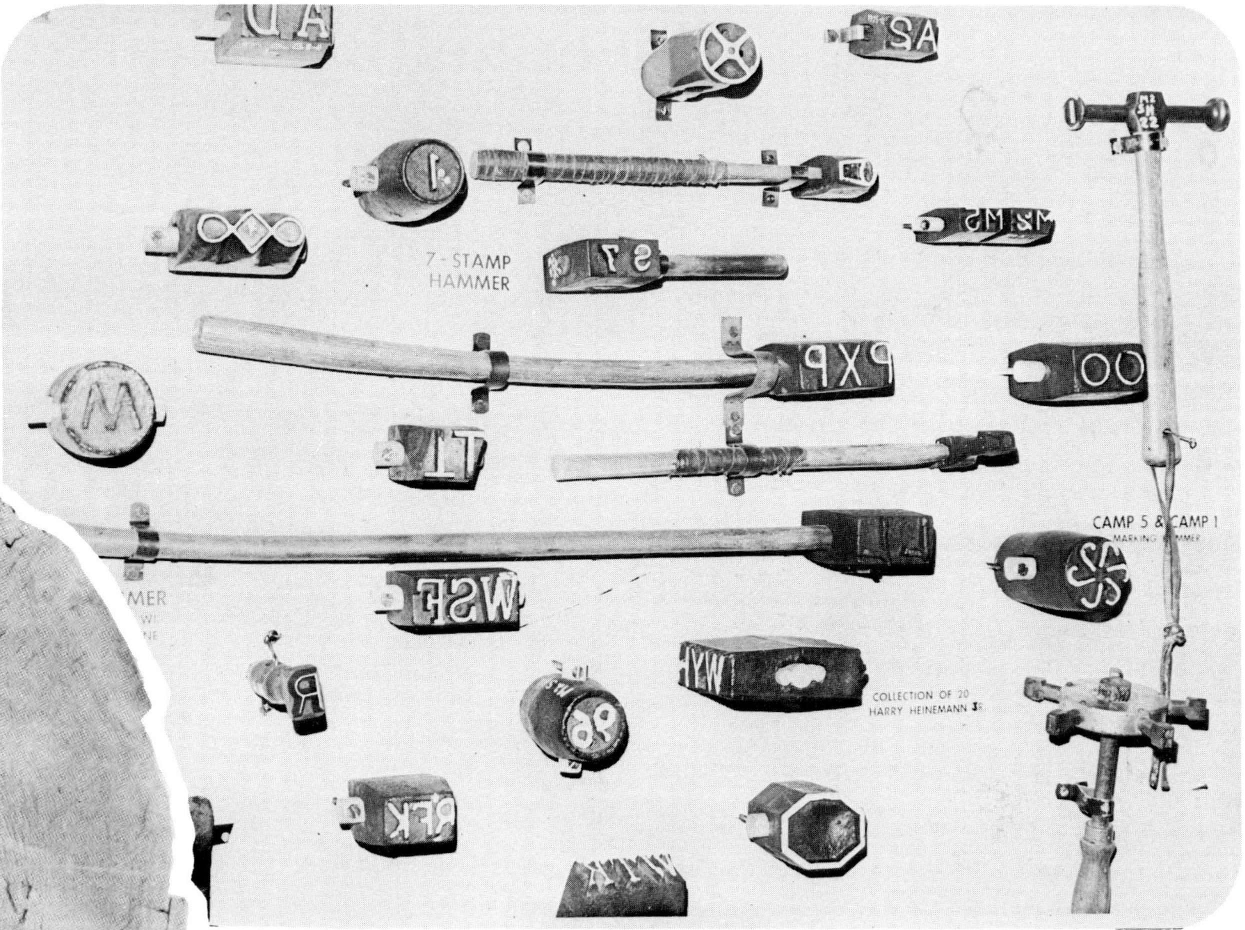

Stamp hammers collected by Harry Heinmann Jr., (Wausau) on loan to Camp 5 Museum at Laona. The reversed lettering has been tinted white to make identification easier.

Stamp hammers from the Chippewa Valley. The top (upper left), bearing the letters HSX, have not been identified. Below, left to right, K for Kaiser Lumber Company; H for Edward Hines Lumber Company; above the H a PC for Dells Pulp & Paper Company of Eau Claire; next right, the circled 43 dash 1, for Christ Muoller, a jobber on the Chippewa River, and above that the PM for L. Schrichn, a jobber, and last on right, LA for George L. Sellers, also a jobber.

A river myth

Old river drivers, often called "river pigs," and lumberjacks as a whole, most of whom have passed to the great lumber camp in the sky, are likely to tell anyone willing to listen that a log travels faster in the river than the current. This belief was carried to the Lake States from the rivers of Maine and New Brunswick, and in Wisconsin it was even given a formal hearing in court, probably in the 1880s, at Oconto. The story is repeated in a privately printed booklet written by William Arthur Holt, a son of DeVillo R. Holt, one of the founders of the Holt & Balcom Lumber Company. The company had made a drive on the Oconto River and a jam had formed which backed up the water to Mrs. (?) Bagley's farm and caused erosion damage. The company offered her $1000 in damages but she refused, and took her case to court.

In testimony heard at the trial, the foreman on the drive said he had raised the sluice gates early to let the water run through the dam for an hour or so before beginning to sluice. He did this, he said, because "logs always ran faster than the water and for that reason it was necessary to let the water run for a while before starting to sluice logs through any dam."

The presiding judge asked the witness how this was possible. According to Mr. Holt, the foreman replied "that the logs were more or less checked by the rapids and by striking the banks and forming small jams, and that a log would run faster in the center of the stream than it would on the sides, and that it was quite possible that what happened was that as he let off an extra head [of water] that morning it had loosened all the logs which had been left hung up on sand bars and along the bank the day before when the head went down, and that these were the logs which reached the Falls [Oconto Falls] sooner than they were expected to."

By that time, the judge was probably just as confused as the witness, but he failed to challenge the myth of the fast-running logs lest he be accused of lack of understanding in the ways that counted along the banks of the Oconto River. Mrs. Bagley got a settlement, but her peers on the jury refused to award her more than what the company had offered her, i.e. $1000. It seems possible that Mr. Holt, writing many years after the fact, misquoted the foreman in some of his testimony, but the fact is that the foreman on the Wolf believed the same myth about a log running faster than the river as Hans Lee on the Little Wolf (page 147). The time allowed for the water to run ahead of the drive, "an hour or so," probably held true for the Wolf, but over on the Little Wolf, a smaller river, this was probably not necessary. Lee said merely that "first you have to let some of the 'head' out before you start the logs through so the 'head' has a good start."

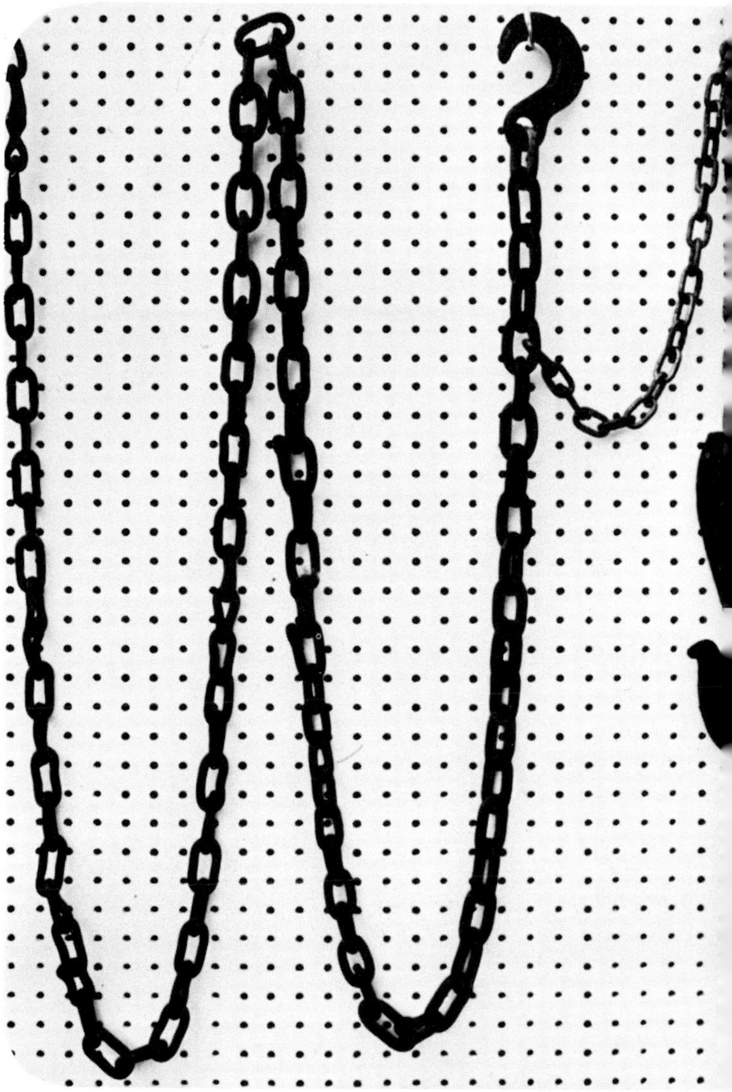

Chain with round hook and fid extension, better known as the "corner bind," this one was made by Gordon Calhouse, blacksmith at Camp 5 Museum.

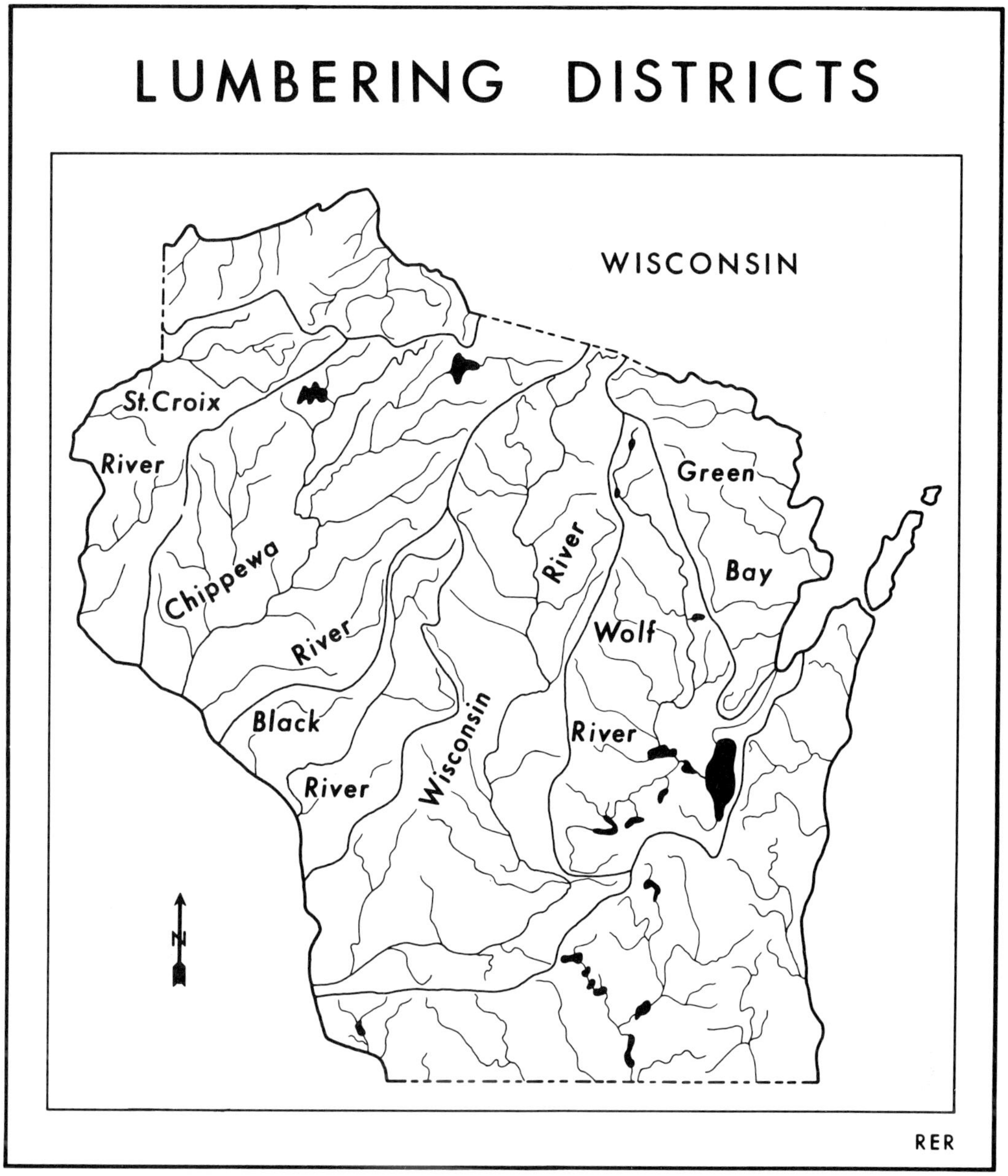

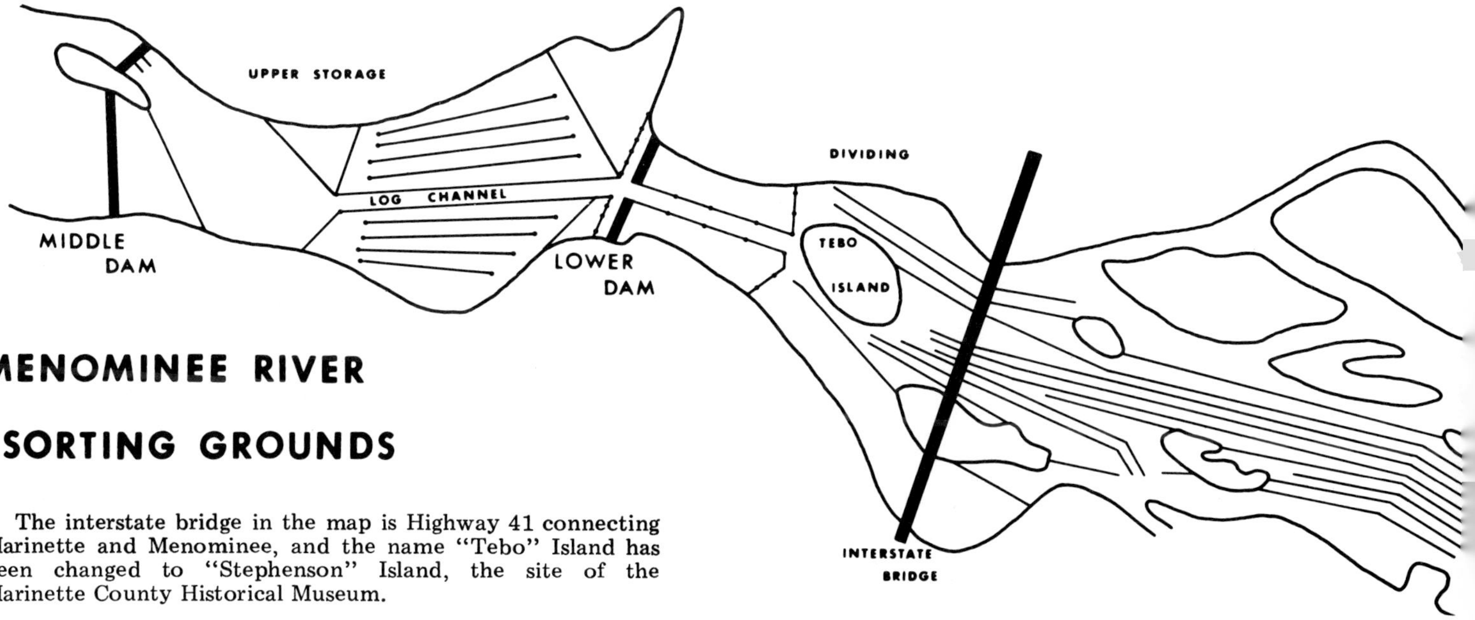

MENOMINEE RIVER

SORTING GROUNDS

The interstate bridge in the map is Highway 41 connecting Marinette and Menominee, and the name "Tebo" Island has been changed to "Stephenson" Island, the site of the Marinette County Historical Museum.

The river rafting days

In the beginning of the lumber industry in Maine, the most important way for lumber to reach a market was to float it down a river. In the beginning of the lumber industry in Wisconsin, there were two ways for lumber to reach a market, either by lake boat on Lake Michigan, or to float it down the Chippewa, Black and Wisconsin rivers into the Mississippi, and from there to Dubuque, Rock Island, Nauvoo, Hannibal and St. Louis.

Lumber rafts, called "cribs," were made up twelve feet wide and sixteen feet long, or they were squared at sixteen feet. These cribs were assembled in sheds near the sawmills and launched into the river on ways. Or, they were made up on the ice, so that they settled into the water when the ice melted. Sawmill owners in Eau Claire and Chippewa Falls had platforms at the tails of their mills on which they built their rafts, and when the raft was ready, the platform was tilted and the raft slid into the water.

There was no rafting above Merrill on the Wisconsin River, but from Merrill on down, through Wausau, Mosinee, Stevens Point and Wisconsin Rapids, millions of feet of lumber passed each year. It was such a common sight in the 1850s and 1860s, the heyday of lumber rafting, that local newspapers scarcely took any notice of the great fleets of lumber moving silently downriver attended by pilots and crews clinging to their long oars, fore and aft, and risking their lives for $1.50 to $3.00 per day.

Rafting on the Wisconsin began in the 1830s and continued into the 1880s by which time the railroads had penetrated the Pinery and made rafting unnecessary, although the last rafts run by the J. Arpin Lumber Company (Wood County) took place in the spring of 1887, and still another company ran the last rafts out of Biron (also Wood County) in the spring of 1888.

The only other big rivers in Wisconsin used for rafting of lumber were the Chippewa and Wolf, but rafting on the Wolf ceased with the coming of the railroad to the Fox River Valley around 1870. On the Chippewa, rafting continued at least another decade, and while some rafts originated at Chippewa Falls, most of them were made up at Eau Claire. Since there were no serious impediments to navigation below Eau Claire, the lumber cribs were often made larger than on the Wisconsin, some sources say sixteen feet wide and thirty-two feet long.

Captain J.M. Turner, writing for the *Wisconsin Magazine of History* (1940), said that the rafts coming off the Chippewa River into the Mississippi were six cribs long and three cribs wide, or eighteen cribs all told. When four Chippewa rafts were yoked together, they made a "Mississippi raft," which was twelve cribs long and six strings wide for a total of seventy-two cribs. But Turner fails to mention the size of a Chippewa crib.

Apparently the majority of the rafts that came down the Chippewa were piloted by Baptiste LaFrance, a French-Canadian whose name had become a byword for rafting on the Chippewa. But the Chippewa raftsmen never suffered the number of drowned and wounded that the raftsmen on the Wisconsin did, especially at Mosinee, better known at the time as "Little Bull" (Falls).

One of the earliest attempts to raft lumber to market was made from a mill built in 1831 by James Lockwood, situated about where the modern city of Menomonie is located on the Red Cedar River. It seems that the French-Canadians working for Lockwood were not experienced in the business of assembling a raft, but they did remember a little about it. The Canadian style, Lockwood says, was to lay two "floats" of ten-inch squared timber down. The lumber was then laid across the squared timbers and bound to the "floats," but Lockwood fails to explain how they were bound or just what he means by "floats." However they were made, this first attempt went amiss when the stage of water on the Cedar River dropped and caused a delay. Shortly later, Lockwood built another mill on a river apparently on Wilson Creek running into the Red Cedar. Here he learned from a Yankee employee how to assemble lumber for a proper "crib" by laying one sawed board lapping about half its length upon another. He got his cribs down the Red Cedar to an eddy where the Cedar joins the Chippewa. Here he was held up by a "freshet" that lasted for nearly two weeks. The water rose so

This illustration of a "Mississippi Raft" is twice enlarged from an engraving appearing in an article titled "Life on the Mississippi" in *Harpers New Monthly Magazine* for December, 1855. But the descriptive material that appears in the article suggests that the author was traveling on the lower Mississippi, probably between St. Louis and New Orleans. The engraving also seems inaccurate. The oars on the raft are too short to do any good, and no Mississippi rafts were made up, all of one piece, as this one suggests. Yet, this small engraving appears to be the earliest available on rafting on the Mississippi River. In the background note the sidewheelers, probably not unlike the steamers piloted by Mark Twain.

high, in fact, that it carried away his mill—the second one in a matter of weeks—and scattered lumber all over the sloughs and bayous of the lower Chippewa River. About 50,000 feet were salvaged and finally run to St. Louis where it was sold at a discount.

When Deputy Surveyor Joshua Hathaway mapped the upper Wisconsin River in 1839, he noted that an eddy below Pointe Basse on the south line of modern Wood County was being used to reassemble and refit lumber rafts which had come through the "Grand Rapids." This seems to be the earliest evidence available on rafting on the upper Wisconsin River, meaning all that part of the river flowing to the north of Portage city.

Henry Merrell, a pioneer lumberman, wrote that he took a fleet of lumber from Portage, down the Wisconsin and into the Mississippi to St. Louis in 1839, but, as noted above, someone else was rafting on the upper Wisconsin earlier, since it is fairly certain that Daniel Whitney, who built the first sawmill on the upper Wisconsin opposite the present city of Nekoosa, circa 1832, must have been rafting lumber to either Portage or to cities on the Mississippi before 1839.

By the mid-1840s, running rafts on the Wisconsin had become such an important factor in the lumber business that something had to be done to assure the raftsmen of the right-of-way through local mill dams that were obstructing navigation. In 1849, the state legislature passed an act (Chapter 62 P.& L.L.) that provided that every dam on the Wisconsin River was to be furnished with a slide or chute for the passage of rafts and watercraft. This slide was to be sixty feet wide; it was to be built of hewn timber, and, in order that the raftsmen would know where the center lay, at least one buoy was to be placed opposite the center of the slide at a distance of twenty yards above the dam. However, no photograph presently available shows that any buoy was ever erected anywhere. It was probably impractical to anchor a buoy this close to a slide. Moreover, it was the business of the pilot on the raft to know where the center of the slide lay.

The technique of transporting lumber downstream in rafts was developed in Canada and by the pioneers of Maine, New York and Pennsylvania. (A model of a Pennsylvania raft may be seen in the accompanying photo.) And the more the "Downeasters" rafted, the greater their self-esteem grew. A story in the Chautauqua County *Democrat* (New York) was picked up by the Wood County *Republican* at "Grand Rapids" on June 16, 1868, and the following exchange was noted:

> The Chautauqua County *Democrat* says a story is going the rounds of the papers of the "largest raft ever known that was recently run from La Crosse, Wisconsin, to St. Louis, which contained a million feet of lumber." We presume a dozen rafts have been floated down the Ohio this spring containing fully that amount. We rode on one that did. Alonzo J. Marsh of this village ran one that contained 1,500,000 feet of boards and carried upon it 1,500,000 shingles. It was 40 rods long and 7 rods wide. Let the Mississippi forest beat that, and then we will tell another.

No matter how large or small the rafts were, most were being constantly challenged by obstacles, visible and invisible, in the big rivers. "Sawyers" were often encountered. These were trees that had been uprooted along the banks and

An artist's conception of a lumber raft as it appears in *Harpers* for March 1860. The oars are incorrectly drawn; the oar locks are too far inboard to be effective, and the raft has no grubstakes to hold it together. The illustration is twice enlarged, and it probably relates to Maine or Connecticut rivers, not to the Lake States.

A model of a Wisconsin River raft, consisting of three "rapids-pieces," was created by an unknown hand in Wausau and is now in the collection of the Marathon County Historical Society. This close-up view of the bow shows how the spring poles were used to raise the bow out of the water. Only two oars are shown, fore and aft; the other oars have been lost, but the oar locks are in place on the headblocks. This model also shows how cribs were yoked together, and how several cribs were yoked together to make a rapids-piece.

had fallen into the river, roots and all, tumbling and rolling with the current.

Channels in the rivers, especially on the Wisconsin, were often changing. But these were lesser dangers compared to the dangers of improperly built, or poorly maintained dams and slides, not to mention the biggest menace of all in the rocky gorges that rushed passed Wausau, Mosinee and "Grand Rapids."

Natural curves in the river, bends and elbows, were added hazards. For example, there was a stretch of water below Nekoosa called the "Grignon Bends," no doubt after Amable Grignon II who once had a trading post nearby. The rafts could not be swung around the bend by the men, wrestling with the long oars, fore and aft. Instead,

Picture at right is alleged to be the last lumber raft on the Susquehanna River (Pennsylvania), and it is used here to show the type of rafts that were built for the slow-moving rivers of Pennsylvania and Ohio as compared to Wisconsin. The oars are just as long as the Wisconsin rafts, but the grubstakes and coupling yokes are reduced because the lower Susquehanna and Ohio rivers are not hampered by gorges. If this was the last raft on the Susquehanna, it was something of a showpiece, too, because the raftsmen are working in white shirts.

A model of a lumber raft on the Maine Line Canal, made by H.R. Rhines for the Carnegie Museum in Pittsburgh and photographed by Robert McCullough and Walter Leuba. This model is probably based on rafts built on the Connecticut River and therefore are smaller than the rafts that came down the Susquehanna and Ohio rivers. The grubstakes are all neatly in place, the cribs are held together with coupling yokes, and the whole is steared by oars. These rafts served as models for the much bigger ones that were to be used in the Lake States.

they had to use the "buck and swing" method. On approaching the first bend, the raft was guided toward the inner bank of the bend, and the bow of the raft allowed to rest on a sandbar or spit of land. The current then swung the stern section around and carried the raft through the rest of the bend, stern foremost. This swinging of the raft, end for end, apparently had to be done two or three times to negotiate the "Grignon Bends."

Another danger spot was at the Dells. Not far above the city of Wisconsin Dells, the river narrows sharply and passes through a rocky gorge, where many rafts broke up. This gorge today is famous for tourist excursion boats.

Below the dam at Kilbourn, many rafts were stranded because the river widens at "Flanner Flats," where crews were apt to wait for days for the stage of water to rise sufficiently to refloat their rafts.

Making of a raft

Several attempts have been made in the past century by former river men and professional writers to describe how a Wisconsin River lumber raft was constructed, but it remains doubtful whether anyone could follow their instructions today and complete a raft that would not fall apart over the first rapids. Photographs of models of rafts, or photographs of rafts taken during the period of lumbering in Wisconsin, give some idea of how the rafts were made, but neither model nor picture can be relied upon entirely.

Perhaps the most lucid account on how a raft or lumber "crib" was constructed is provided by Adolph Kenneberg, counsel for the Public Service Commission of Wisconsin who, in 1944, submitted a report on "log Driving and the Rafting of Lumber in Wisconsin." The survey was mainly concerned with the question of navigation on Wisconsin waters, but it also dips into other facets of information on log driving and rafting, and in one of the chapters, Kenneberg describes the making of a raft:

> A Wisconsin lumber raft was composed of what were called cribs. Six cribs yoked together constituted a rapids-piece. Three rapids-pieces placed side by side made up a Wisconsin raft. The cribs were either 16x12 feet or 16 feet square. Each crib contained from 12 to 20 tiers of inch boards, depending upon whether the lumber was green or seasoned and the depth to which it was safe to let the crib sink into the water without taking too much risk having it wrecked upon the rocks in the bed of the river.
>
> In making up a crib three 16-foot grub planks 2 inches thick were laid parallel to each other and parallel to the stream. Three 2-inch auger holes were bored in each grub plank, one at each end and one in the middle. Grubstakes 2 inches thick and about 5 feet long having a knob at the lower end were inserted into the auger holes from the underside so that they stood up about 3 feet above the grub plank. Three 3-inch cross planks with auger holes was [sic] laid across the grub planks and held in place by the grubstakes which passed through the auger holes The grub planks and cross planks formed the foundation of the crib.
>
> First a layer of 1-inch boards were [sic] laid upon the cross planks. The next layer was laid crosswise upon the first, and in a similar manner each succeeding layer was laid crosswise upon the preceding one until the pile was completed. Two binding planks 16 feet long with 2-inch auger holes corresponding to the holes in the grub planks were laid in [like] manner to receive the grubstakes. The grubstakes were pulled up by means of a lever and chain called a 'witch' whereby the lumber became bound firmly between the grub planks and the binding planks. The grubstakes were then fastened securely with wedges and the crib was completed. Cribs were placed end to end and yoked together with 20-inch planks with auger holes at their ends. The yoke planks were placed over the end grubstakes between the cribs, thereby forming a flexible coupling . . . .

Rafts were guided by long oars, forty-five to fifty feet long. The oars were made of planks, sixteen feet long, coupled to a long, heavy pole smoothed off at one end for easier handling. The oar was balanced at the bow and stern of the rapids-piece on a timber, called the "headblock," which was six by eight inches in thickness. The headblock was wedged securely into the raft by grubstakes, and a grubstake in the center of the headblock acted as oar lock. The men handling the oars at the stern of the rapids-piece were called "tailsmen," and the men up front, "bowsmen."

The final touch in the construction of a rapids-piece was the insertion of "spring poles" in the bow, that is, in the front end. These heavy poles of tamarack, twenty-five to thirty feet long, tapered twelve to fourteen inches at the butt and six to eight inches at the top end, were inserted under the headblock of the foremost rapids-piece, and, this done, eight to ten men stood on one end of the pole to force it down to where it could be secured to a grubstake. A hole was augered into the end of the pole to match the width of the grubstake situated on the side of the piece. Two or three of these spring poles then acted as a spring which lifted the leading edge of the crib partly out of the water to make a curved ski-tip prow. This improved navigation and the raft was less apt to catch on snags and sandbars.

The grubstakes mentioned above, often called grub pins, or plain "grubs," were small saplings pulled up by the roots in the forest and trimmed to four feet in length from the root. The stake was shaved all the way down to the root so that holes in the planks fitted snugly over the stakes. The

There are two sawmills in this picture taken on the Chippewa River at Eau Claire, and standing between them is an unusual sight: a shed of the Daniel Shaw Lumber Company where rafts were "rafted," that is, assembled. (The verb "to raft" was used both to make up a crib of lumber, and also to take the lumber downriver.) In this picture, lumber cribs are being assembled inside the open-air shed and slid down the ways into the river. The rafts shown on both sides of the river are also larger than those found on the Wisconsin River, and less attention has been paid to coupling yokes and grubstakes. There were no rapids to negotiate on the lower Chippewa, that is, below Eau Claire, and for this reason the so-called "Chippewa raft" was larger and less securely put together. Note the oars on the raft at the right. The small pieces on board the raft are "deck load," apparently lath. At the left is the one-time sawmill of the Empire Lumber Company of Orrin H. Ingram, a pioneer to Eau Claire who succeeded where others failed.

root was also trimmed to within an inch or so of the bottom to make a knob just big enough to act like the head of a holding bolt. In this manner, the stake could not be pulled through the lumber crib from the top. At the top of the stake, a wedge, as Kennenberg mentions, was driven into the side to prevent it from sliding in the opposite direction, that is, downward through the planking.

Thousands of these grubstakes were needed by the lumbermen every year, and many pioneer farmers spent their spare time in the fall searching for saplings, pulling them up, no doubt with the help of oxen or horses, and hauling them home where members of the family worked on them through the winter days and nights. Grubstakes could be used for legal tender and they could be exchanged at the local store for groceries—or whisky. In the spring of 1864 grubs were $35 to $40 per thousand. In the spring of 1872, the Wausau *Pilot* was complaining about the high price of grubstakes, which were selling at seven to eight dollars a hundred! Yet, the newspaper admitted that the lumbermen around Wausau needed 15,000 grubs that spring.

So many of these grubs were needed that a special storage shed was built by mill men to store them until needed. And just about the time when grubstakes were no longer needed, someone invented a machine to turn them out on a lathe!

The earliest description of rafting lumber on the

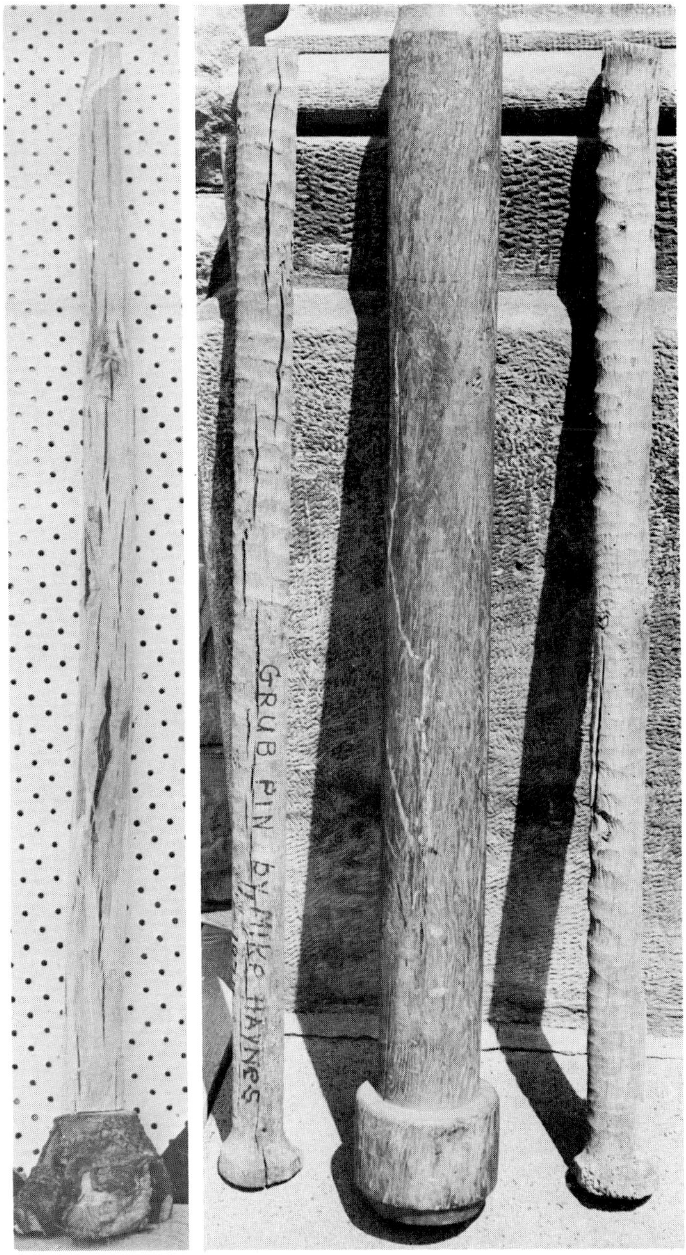

Grubstakes were used to bind lumber and planks in a raft together by boring a two-inch hole through the lumber, top to bottom, and inserting the grub from the under side, leaving the root at the bottom. Most grubs were handmade from green saplings and whittled to size with a drawshave, as seen in photo at the left. This one, made by the author, is four feet long, the standard length for Wisconsin River rafts. In the 1870s, just when rafting was coming to an end, grubs were made on a lathe (right). The two narrow ones were found by Mike Haynes and Frederick Borm in the Red Cedar River, below Downsville, and each is forty-four inches long. The large grub in the center is a freak. It was taken from the mill pond at Menomonie and was probably used on a key boom stick to open or close the log-holding boom.

Chippewa River was written in 1875 by Thomas E. Randall, a local historian from Chippewa Falls. His description reads, in part, as follows:

> The method or art of rafting lumber on the Chippewa is very much like that pursued on the Delaware, Susquehanna, Allegheny and Ohio rivers. Cribs sixteen by thirty-two feet, instead of the old style platform, is now the invariable shape of the first compact form of raft into which the lumber is placed and firmly fastened by ironwood 'grub' pins inserted through two-inch plank, those on the under side or bottom of the crib being called 'runners' and those on the top 'binders.'
>
> The grub pins are turned in a lathe [this is the first evidence that the homemade grubstake was being replaced by a factory-made grub] with a head on one end which is neatly fitted into the underside of the lower plank or runner, so as to make a smooth surface.
>
> These cribs are formed on ways, with rollers on which they rest, and when the crib is of the desired thickness—twelve to twenty courses according to the stage of water—a strong lever called a 'witch,' having a heavy iron clamp on one end, is placed over the end of each grub pin, with a fulcrum bearing on the binder, by which means the grub pin is drawn upward, thereby causing the runner and binder to be drawn together with great force, binding the lumber between tightly together, when the grub pin being securely wedged, the crib is finished. The fastening which holds it to the ways being now removed, the crib slides into the water. Seven of these cribs constitute a 'string' and four strings form a Chippewa raft . . . .

On the Wisconsin River, the "strings" referred to above became "rapids-pieces." But only three "strings" or rapids-pieces constituted a raft on the Wisconsin.

After running the falls and dams at "Grand Rapids," the narrow gorge at the Dells, and the dam at Kilbourn, the rapids-pieces were linked together for the last time on the first eddy below the Dells and from there on it was fairly uneventful all the way to the Mississippi unless the pilot forgot where the main channel was, and this could happen to the best pilot in the shifting sandbars of the lower Wisconsin.

When all was going well, meaning that the stage of water was high, but not too high, a rapids-piece leaving Wausau could reach St. Louis in twenty-four days. In some seasons, when the stage of water was especially favorable, it was possible

A scene on the Wisconsin River at Wausau probably copied from a tintype or wet plate. The copier has inscribed the date as 1865. The scene suggests how cribs of lumber were being "rafted" that is, assembled, into rapids-pieces in preparation for the first leg of a journey that would probably terminate at St. Louis, MO. A wing dam has been built at right center to allow for a slack water where the cribs could be assembled. In the distance is the covered bridge that replaced an earlier uncovered bridge, both predecessors of the present Scott Street bridge in Wausau.

Right: dam and sluiceways probably on Wisconsin River in Wood County, sawmill on opposite bank. At left, three men in stovepipe hats study the dam. A woman with dog walks at the right. The river is in flood tide and the picture, no doubt a wet plate, was probably taken in the 1860s.

for a crew to make three round trips, but most of the time two trips were considered normal, one in April and one in July. A dry spell, with no rain to sustain the level of the river, was more feared by owners than the rocky gorge at "Little Bull." A fleet of lumber stranded on a shallow, or sandbar, could become so solidly mired in mud that it could not be refloated, and in many cases it cost more to retrieve it than to abandon it.

The delays caused by low water could even provoke humor. In 1840, according to oral tradition, a Mr. Barnes left Pointe Basse with a fleet of lumber bound for St. Louis. Owing to the low stage of water, it took him so long to reach Kilbourn that he ran out of food for the crew. He planted potatoes near the Dells Eddy, and on the next stretch of the river, it took him so long to reach Sauk Prairie, sixty miles downstream, that he ran out of food again. A detail was then sent back to the Dells to dig up the potatoes!

Rafting lumber was a financial risk, and the wonder of it is that more men did not lose their fortunes in it than is generally reported. But, by the time that all the risks had been taken, it was mostly companies with big capital investments who survived. They could afford to lose rapids-pieces, or even a fleet of lumber. Moreover, there was no danger in those days of being sued by the family of a man who lost his grip on the "sucker line" and washed overboard at "Little Bull" or on the dam at Kilbourn. (The sucker line was a rope stretching along the deck of a rapids-piece from bow to stern to which the men clung when they went over a dam or through a rapids.)

Every rafting fleet had a pilot under whom the several raftsmen took orders. The pilot was capable of negotiating any rapids or dam on the river, but he usually deferred to a local pilot to take a rapids-piece through the local falls or over a dam. One of these skilled men was William Curer of Mosinee, well-remembered by Edgar Ladue in his book, *Early and Late Mosinee* (1907). Said he:

> . . . Mr. Curer . . . sees the boy he has engaged to handle his horses and says for him to be on hand bright and early in the morning, and at daybreak you would see Curer on one horse, the boy on the other, making for the Eddy (above Mosinee) and on reaching there, Curer jumps from his horse and runs

onto the string of rafts that has been prepared for him, and [starts] for the gorge. The boy has started already with both horses for the potato patch below the sawmill, where he is to wait until Curer is taken from the raft in a skiff and brought to shore, lights his pipe, jumps onto his horse and flies for the Eddy as fast as his horse can go . . . jumps onto another rapids-piece and away he goes again. He always runs with two crews, one going, the other coming all day long. I have known him to make thirty trips in one day. He always had one dollar for each trip through Little Bull.

. . . I have seen a raft of ten men, five to each oar, every man of them all out of sight under the water at the same time, not a particle of the raft or men could be seen for a few seconds, but several hats are seen floating on the bellows around. At such times of running the falls, everybody who could, men, women and children would hurry to the falls and stand around or sit on the rocks and watch the rafts as they came down the slide into the gorge. It was a gala day for Mosinee. There were hundreds of raftsmen in and around Mosinee at such times . . . going through [the falls] and coming back all day.

Faced with the several dangerous rapids and dams on the Wisconsin River, raftsmen risked death in a casual manner. It was the price for making a living in those days. In one instance, at least, it is recalled that the men of one crew at Mosinee held a brief prayer meeting before tackling Kelly Rock in the center of the Mosinee gorge.

"I remember well," said Ladue, "it was on the 23rd day of April 1874, that I was on my way to the city of Wausau, and at Cedar Creek I met Mr. Michael Stafford. We stopped and talked for a few minutes and he told me he was on his way to Little Bull to help the boys run his lumber over the falls. He went on to Mosinee and I went to Wausau. When I returned home I heard the sad news that Mike Stafford and four other men had been drowned at the dam below the falls . . ."

Figures on the number of men who lost their lives on the river are bound to be suspect and unvarifiable. *The Central Wisconsin* (Wausau) reported that in 1872, twenty-seven men were drowned at the Clinton dam below Nekoosa. No one ever added up the total fatalities for the entire river that year, or for the years ahead.

A few years earlier than the above story, an advertisement ran in the Wisconsin *Lumberman* (May 12, 1865), which had this to say: "[There are] rumors that the Edwards & Clinton "slide" is not safe to run lumber over. But I do aver, that there is no danger in running lumber through the slide or over the apron and this is attested to by the undersigned pilots who are acquainted with the slide and have been running it for years and are all known to be First Class Pilots."

John Edwards, who paid for this advertisement, said if anyone thought the slide was not safe, he would "insure each rapids-piece at [for?] twenty-five cents," provided it was run by any one of the pilots named below: Joseph L. Whitney, F.B. Case, H. Canning, John L. Lavine, Ole Natwig, Louis Grignon, Paul LaRose, Charles Richards, R. Harvey, Joe Barshaw, B. Buck, and Tim Bacon.

Obviously, the Clinton dam must have deteriorated between the time this advertisement appeared in 1865, and 1872, when twenty-seven men lost their lives there, and Edwards either had gone down himself or was keeping mum about it because no further advertisements appeared on the safety of the Clinton dam.

Before it was reconstructed in the early 1860s, the dam at Kilbourn was a death trap which claimed the lives of many more men, and smashed up more rafts, apparently, than any other on the Wisconsin River. Finally, lumbermen from Grand Rapids, Stevens Point and Wausau, led by Walter McIndoe, raised a fund to fight the village of Kilbourn through the courts, at which time the village agreed to rebuild the dam.

Newspapers of the period, all weeklies in northern Wisconsin, did not carry much news about the great rafting epoch that was passing them by. Most of the weeklies were owned and operated by one man, probably assisted by his wife, and the editor did not have time to run out and "cover" a story, no matter how important or tragic. Most news items were brought to him and if no one mentioned that a man was overboard in Conant Rapids, there was no news about him. For this reason, probably, a search through the newspapers of Grand Rapids, Stevens Point, and Wausau reveals few names of men who gave their lives to bring lumber and wood products to the frontier towns of America.

The rapids-pieces shown here are part of the last fleet of lumber run for the J. Arpin Lumber Company (Wood County) in 1887. H. (for Hamilton) H. Bennett, a pioneer photographer of Kilbourn (Wisconsin Dells), must have realized the nature of the occasion and joined the fleet at some point up the river, probably above the Dells. He was probably standing on a bridge or cliff to get this shot, just below Kilbourn where several rapids-pieces are being assembled into "Wisconsin rafts" after having come through the Dells and the several "elbows" and hazardous turns that marked the passage through the Dells. The rapids-piece in the foreground has a "kennel," or small house which could sleep two of the crew. The long handle of one oar forms a shadow in front of the man at the left, and at the rear, the men are preparing to couple up two rapids-pieces.

The risks they took

One of the rare accounts of rafting from the early period on the Wisconsin River appears in the *Pinery* at Stevens Point on May 3, 1855, written by someone who signed only his initials, "C.E.", but who was, no doubt, better educated than the average raftsman. He describes his experience:

> Peter Grover and Wm. Loing had taken the contract to run all the lumber for the Little Bull Company this season, including in all I think about two million feet. [They] have between 40 and 50 choice raftsmen at work now, putting cribs into the water, coupling up, etc., and the way the boards rattle, axes hack, and the lumber drops down the stream is evidence of what men can do when in vigorous health, with willing hands, good quarters, warm blankets and plenty to eat. . . .

By "Little Bull Company," the correspondent no doubt refers to the firm of Joseph Dessert and associates at Mosinee who, in 1849, took over the original up-and-down mill built by John L. Moore in 1839-40. Grover and Loing, both pioneers of Amherst, were not part of the Little Bull Company, but had contracted to deliver their lumber to markets on the Mississippi River.

Continuing, the letter to the editor of the *Pinery* had this to say in part:

> The secret of Peter Grover always having so good hands, is to be found in these simple facts: he always pays all he agrees to—something I wish I could say of more persons than circumstances will permit of; he goes in and works equally as hard, if not harder than any of his crew, provides well, is wide awake while others sleep, and[is] good natured . . . .
>
> Yesterday the river was crowded with ice, rushing down in cakes of all sizes, and the water is rising rapidly. Little Bull Falls presents an appearance not calculated to quiet the nerves of the timid; water, ice, logs and all kinds of driftwood come rushing in, grinding and crackling, and as it gets seemingly on a good course down the river, it wheels round to the left and marches straight into the jaws of the falls . . . . These falls have the reputation of being the most dangerous on the river. A good supply of shingles for loading are scattered along the banks, which must be put upon the rafts, and then we shall be ready to drop down with our fleet . . . . But more lumber will go out of the river this season than last, by fifty percent.

On May 10, 1855, the *Pinery* reported that rafting had about eight days of the "best kind of running [and] got out all the lumber with great rapidity. The water went down suddenly on the 6th and 7th [of May] and a general suspension is the consequence. Messrs. McIndoe and Shuter [Wausau] had at least $100,000 worth of lumber on the river [in Stevens Point] waiting for a shower to raise the river."

Here is an example of the risks that the lumbermen faced every time they sent a fleet of lumber down river. McIndoe was at the mercy of a rain shower to save $100,000 worth of lumber!

The same issue of the *Pinery* reported that at Conant's Rapids (south of the present Clark Street Bridge in Stevens Point) Sam Harlow was "lost" while attempting to ride a skiff to a crib of lumber that got fast in the river just above what is known as the "red rock." Two other persons were in the boat when it capsized, William Reynolds and Samuel Jackson. Reynolds got onto the crib and Jackson swam to a rock and was saved.

A few weeks later, the *Pinery* (June 28, 1855) reported that John Gardner of Stevens Point was at Little Bull Falls where he was attempting to snub a raft to another raft with a cable by passing the line around one of the grubstakes. As the cable was paying out, his foot became "embraced in the coil, and was almost instantly crushed against the grub, the flesh being torn off and the bones broken."

The victim was not rushed to a doctor. He was carried, probably, in the back end of a wagon, and by the time he reached the doctor, he died.

The Wood County *Republican* (Grand Rapids) carried only a few items on the raftsmen in more than three decades of publishing. On March 26, 1859 it carried a report clipped from the Kilbourn *Mirror* which had this to say:

> The first raft went over the shute [sic] of our

→

This is perhaps the most striking picture ever taken of rafting in Wisconsin, another in a series by H.H. Bennett, in 1887. Stop-action photography was invented in the 1870s, but it is believed Bennett had devised a shutter of his own to stop the action. The picture shows a rapids-piece, nearly submerged, diving over a rough spot in the Dells while the oarsman struggles to keep the piece from smashing against the cliff at the right. A manmade barrier is visible below the cliff where, no doubt, other rafts have foundered. The picture also reveals how a rapids-piece could bend, since it seems certain that Bennett was standing on a level higher than the bowsman. Another oar lay astern but this could not be photographed from his angle. Note the grubstakes and coupling yokes in center and along the sides, and, up front, the spring poles on either side to raise the bow a little out of the water when running on even keel.

The dam at Kilbourn in the 1850s was a death trap where many raftsmen lost their lives, but after 1860, it was less a hazard after the village agreed to lower the dam and improve the slide through the center. Here a rapids-piece is coming over the slide and two men, in white shirts and vests—no doubt passengers—are taking the dive standing up for the benefit of the cameraman.

[Kilbourn] dam last Friday afternoon, piloted by Patrick O'Hare of this place, and the way Pat leaped and shouted and cheered as he was going over, helped the fun and excitement of the hundred spectators on shore decidedly."

Pat was no doubt quite aware of his audience as he plunged over the Kilbourn dam where so many men had already been lost, and he wanted to give the spectators a good show by shouting probably to keep his own fear from showing.

The following year, the same paper in "Grand Rapids" (March 17, 1860) reported that the city was waiting for the river ice to break up at which time "the river will be free from ice and our streets thronged with red-shirted pedestrians. . . ."

In the May 14, 1855 issue of the *Pinery*, the editor, A.G. Ellis, reprinted a letter written from "Grand Rapids," but which was apparently mailed from Little Bull Falls and sent to the editor of the *Spirit*, the first newspaper in Waupaca. The correspondent had this to say about his experiences as a raftsman:

> After eight or ten days of hard rafting, the long hoped for sound of 'tie loose' from our pilot, caused a spasmodic movement of the boys, and ever since then the oars have kept up a continual creaking, and the excitement has been intense. In one bend of the river, when the eye could get [see?] no more than half a mile range each way, I counted 22 Wisconsin rafts under motion. At every feasible tying up place we found rafts piled in, filling up all the eddys [sic] and occupying all the snubbing posts."

Farther on in his letter, the same correspondent reported that "today an old timber raft, owned by J. Beedle, went to pieces here in Grand Rapids and six men got overboard. A young man by the name of Hugh Burns was picked up at the foot of the rapids, nearly killed, but it is hoped he will recover. Three others came near drowning but were rescued."

In the above, the correspondent says "after eight or ten days of hard rafting." Here he is referring to the act of making up the rafts at Mosinee, since the order to "tie loose" comes after, not before the rafts were made up. Thus it seems that the term "rafting" was used both in the preparatory stages of making a raft, and later while running the river. The word "sailing" was not used, of course, because there were no sails on a Wisconsin raft.

The letter to the Waupaca *Spirit* also mentions "snubbing posts." These could be trees on shore, or a big stump, or piling that had been driven into the river bed for the express purpose of using it as a snubbing post.

Pilot fees and charges

The cost of running a rapids-piece over the dam at Stevens Point is given in an advertisement in the *Pinery* for March 19, 1857. Larry Gates, a pilot residing in the city, said he would take a rapids-piece over the Conant Dam slide for $1.00, but he made no guarantee that he would get the piece through at this price. For $10 he "warranteed" safe passage for each piece per passage. A decade later William Curer was taking a rapids-piece through a much more dangerous falls at Mosinee for $1, almost surely the result of experience and skill accumulated over the years.

The lumbermen who sent their rafts downriver either used their own crews, or hired contractors, as already noted, who specialized in the business of rafting. In many cases, the pilot who went with the fleet to market in Dubuque or St. Louis was trusted to bring back the proceeds. It was not easy to transfer funds in those days and a courier was often used. The rafts were not always consigned to one customer. The pilot might swing his "Mississippi raft" into Dubuque first and drop a string of rapids-pieces, and repeat the process in the next river town downstream before reaching St. Louis.

Before 1870, the trip back to Wausau, for example, was made by river steamer from St. Louis to Galena, or up the Wisconsin River as far as Portage, and from there the men either walked or took the stage to Stevens Point and Wausau. Most of them walked. The price of a ticket from Stevens Point to Wausau was six dollars and that was more than a common raftsman made in three or four days of work.

Local supply firms advertised their wares for raftsmen. An advertisement in the *Pinery* (April 5, 1867) for Krembs & Bros. Hardware boasted "the best stoves for raftsmen ever brought on the market [and] made on purpose for the river trade. . . ." William Clifford, also a merchant in Stevens Point, advertised (February 18, 1869) that he had ten thousand rock elm "grubs" for sale and he "warranteed" them, too.

Some idea of how rapidly the business of rafting had grown in fifteen years since it began on the Wisconsin is suggested by an advertisement appearing in the *Pinery* on April 3, 1855, in which

Not many raftsmen owned a pocket watch, but this one, probably a pilot, was careful to leave his watch fob showing, as well as his suspenders. He struck the pose for Bennett when the Arpin fleet passed through the Dells in 1887. The picture also gives a close-up of the shaft holding one of the oar blades. Although the oar rests on an even keel in the picture, when the blade is dipped into the river the shaft will be as high as the man's head, and the oarsmen will reach up with both hands and swing it back and forth. It was never called a tiller.

the sawmill firms of Merrill & Slothower (Stevens Point) and McIndoe & Shuter (Wausau) were calling for one thousand men "this spring to raft and run lumber to market. Liberal wages from $1 to $3 per day."

These wages were probably not quite as liberal two years later when the financial panic of 1857 hit the nation. In fact, the setback affected the lumber industry, in camp and sawmill, and it was not until the great demands of the Civil War were felt that lumbering again came into its own. And then the economy boomed. While President Lincoln was calling for more volunteers, mill owners on the upper Wisconsin were calling for men to come and work in the woods, in the mills and on the rivers. An advertisement in the Wisconsin *Lumberman* (March 23, 1864), signed by the leading lumbermen of Stevens Point and Wausau, called for five thousand men. In fact the logging woods, where no questions were asked, became a haven for draft dodgers and deserters. The demands for wood products used in the building of army camps, gun emplacements, and railroads were insatiable.

A week later the *Pinery* at Stevens Point was running an advertisement for three hundred men to run the river for Clements & Walker who offered $1.50 to $3.00 per day.

But, with the coming of the railroad to Stevens Point in 1871, and to "Grand Rapids" and Wausau shortly after, the death knell of the great rafting era was sounded. The last fleet of lumber for the J. Arpin Lumber Company (Wood County) was run from "Grand Rapids" in 1887. The late Gustave A. Giese of Wisconsin Rapids wrote an account of his rafting days on the Wisconsin under date of April 3, 1947. This was later given by his daughter, Mrs. Fred Kester, to L.G. Sorden (Madison). In his account, Giese says, without hesitation, that he was "the last survivor of those that went down the Wisconsin River on the last fleet of Daly & Sampson" in 1888. The fleet originated at Biron (Wood County) and was piloted by John Parrish, no doubt one of the last of the great river pilots who survived at least forty years of rafting.

Mr. Giese remembered that the lumber cribs in his day were sixteen feet square with twenty-four courses of one inch stuff totalling 6,144 board feet of lumber per crib. He seemed unaware that earlier cribs were twelve feet wide and sixteen feet long, but he confirmed the general rule that seven cribs placed in tandem made a rapids-piece and three rapids-pieces coupled side by side made a "Wisconsin raft." He was also best acquainted with the Wisconsin River below Biron, not above that point. There were two dams to run, one at "Grand Rapids," called the "South side dam," and another below Nekoosa called the Clinton Dam, already mentioned. Raftsmen running this stretch of the river tied up at Four mile Creek and "gigged" back, meaning they either walked or rode back to the spot where more rapids-pieces were waiting to be run over the two dams.

The Wisconsin rafts were assembled at a landing near Four Mile Creek, according to Mr. Giese. The coupling yokes used to bind the rapids-piece together to make a Wisconsin raft were made of jack pine about four feet long and six inches thick. These coupling yokes were slipped over the grubs from one rapids-piece to the other to bind them together.

While the Wisconsin rafts were being put together at Four Mile Creek, mill owners were bringing down "deckload," that is, shingles, laths, fence posts, short lumber, timber (i.e. square timber), and extra coupling yokes, to be carried on the decks of the rafts. The pilot always had to keep a sharp eye on the shingle bundles which could be exchanged ashore, at night, for a pint of whisky.

The landing at Four Mile Creek was also the place where small cabins, not much bigger than a dog house—and called "kennels"—were built of rough lumber, but each was big enough to sleep two men. In addition to the "kennels," a lean-to or roof of some kind was built on board one of the rafts to enclose, at least on three sides, a place for the cook to prepare meals for the crew. The raft that carried the cook and food supplies was, of course, called the "G.H.Q." or the "cookraft." The men ate their breakfast and late supper on the cookraft, skiff men carrying them from the other rafts to the cookraft and back, until everyone was fed. Noon meals were delivered to the rafts by the skiff men because there was no time for the men to

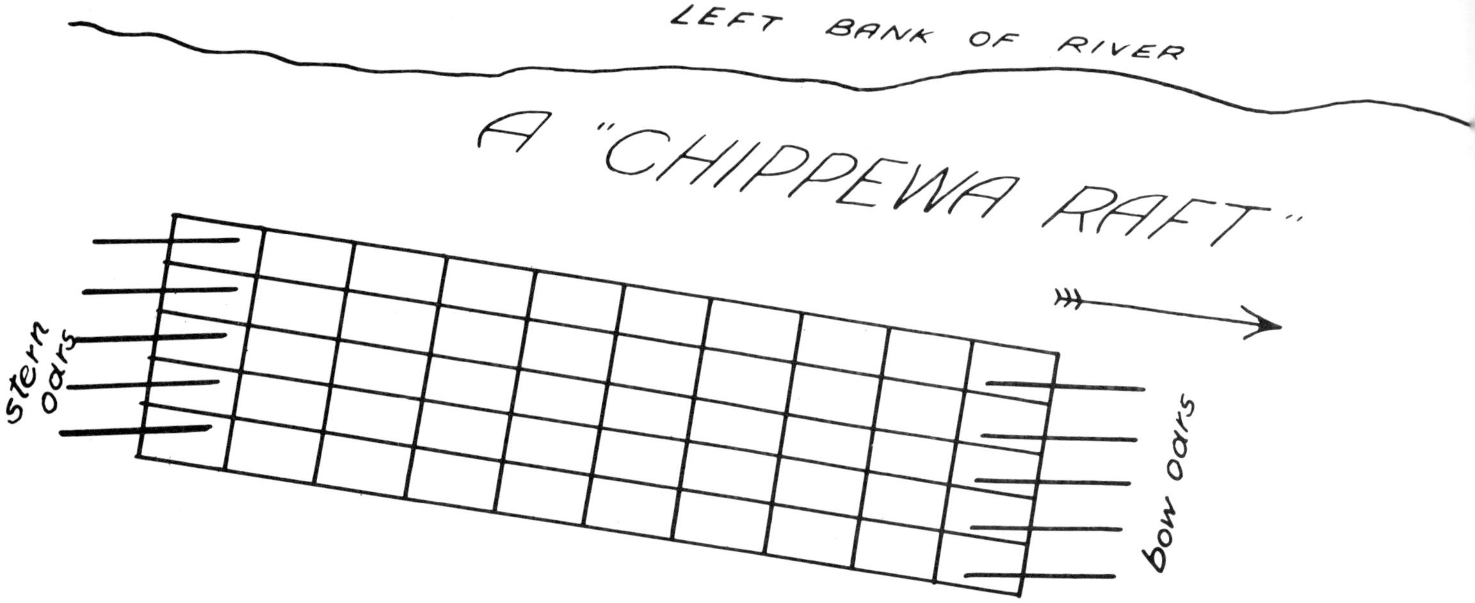

A sketch of a lumber raft on the Chippewa River, done by an anonymous artist, shows fifty cribs, ten long and five wide. The Chippewa raft could be anywhere from two hundred to three hundred feet long, and eighty-five feet wide. In this sketch the artist explains that the bow of the raft is being moved to the right, the stern to the left, called "cross pulling." The raft will make its own way toward the right bank of the river in this instance.

Lumberjacks traveling between job opportunities in Eau Claire and Black River Falls often used this stage coach operated by William Price. The signboard at the right advertises apples, flour and feed. Cartoon art decorates the door of the stage. Note long hand brake at driver's right.

leave their posts on the long oars, fore and aft.

After leaving the Wisconsin River and entering upon the Mississippi, there was no stopping at night to go ashore for a drink or good time. Here all the Wisconsin rafts were coupled together to make one large Mississippi raft. Coupling yokes, similar to the yokes used to tie rapids-pieces together, were also used to tie Wisconsin rafts together to make a Mississippi raft, which, according to Giese, could be one hundred and sixteen feet wide and six hundred feet long.

Since there were no "Grignon Bends" to navigate, or dams to run over, the crews on the Mississippi were small. Often the big raft tied up at Dubuque, and the pilot paid off at least half of the crew. These men either returned to their homes in Wisconsin, or went into the Dakotas to work in the wheat fields until fall before returning to the woods in Wisconsin.

One of the main obstacles to navigation on the Mississippi, aside from sandbars and "sawyers," was the wind. A strong wind could blow a big raft off course, to one shore or the other, and there was little that the crew could do about it except wait until the wind changed or died down. Geise said it was hot in the day and cold at night on the river. The men were told they could drink the water out of the Wisconsin, but not out of the Mississippi. The cook had two big drums filled with hot tea or coffee for the men to refresh themselves. But the work was easy and a relief from the strenuous days that had carried them over the dams and through the rapids all the way from Big Bull Falls (Wausau) to the Dells of the Wisconsin River.

According to Donald Fleutsch (Cassville), a "ginpole" was used to swing rafts around short bends and sandbars on the Mississippi River. This was in the days when rafts were small and before paddle steamers and bowboats were brought in to handle the larger rafts or brails which were being assembled in the 1870s. The ginpole was about twelve feet long, pointed sharply at one end. Three feet or so from the point was an eyebolt and ring to which was fastened the check line holding the raft. Two men took this pole and check line in a skiff and rowed to a sandbar where they tilted the ginpole into the sand at an angle of about forty-five degrees. The pull from the raft then forced the pole into the sand, and it acted as a snubbing post. The pole was removed from the bar merely by pushing the tip upward at which time the check line to the raft was released, or unhooked.

There is no explanation for calling this snubbing device a ginpole, which ordinarily denotes something quite different in lumber operations.

Although the Wolf River is not associated with the rafting of lumber, it seems clear that before the railroad came to Shawano County in the 1870s, rafting was the only means of sending lumber to cities downstream and to Lake Winnebago. On May 18, 1849, the Oshkosh *True Democrat* reported that Mr. E. Gorden "has arrived here during the few days past with one and a half million feet of lumber from his mills on Lake Shaw-anow and Little Wolf. Besides this, others are bringing, or preparing to bring lumber here. Lumber yards are started and the appearance is of doing a large business."

But rafting of lumber was overshadowed on the Wolf by the log drives which were made to the sawmills at Shawano, Oshkosh and Fond du Lac. Here the logs were sawed and shipped out by train to wholesale yards in Milwaukee and Chicago.

With the coming of the railroad to Stevens Point in 1871, and to other cities in the Pinery by 1880, the rafting picture also changed in north central Wisconsin. Lumber could be sawed and dressed in Stevens Point, Chippewa Falls, Black River Falls and Menomonie and shipped out by train to market, not only on the Mississippi but throughout the country.

But old ways die slowly. The last lumber raft on the Mississippi was made up in Stillwater, Minnesota, on the St. Croix, and left that city on July 29, 1925, bound for Fort Madison, Iowa. *Ottumwa Belle*, Captain Charles Brown, had charge of the raft and *Pathfinder* was the bowboat.

And then the order to "tie loose" was heard no more from the river men.

Another Bennett picture, this one, no doubt, taken above or below the Dells, shows a crew of men using hand spikes to shove a rapids-piece off a sandbar. The grubstakes and coupling yokes holding the several cribs together are plainly visible. On board at right is "deck load," all lath.

The *Annie Girdon*, "Lightning Packet of **Menomonie," nudges a raft into the river, probably at the mouth of the** Chippewa River on the Mississippi, for Knapp, **Stout & Company**. She was built at Burlington, IA in 1866, forty-six tons, **with two engines and one boiler. W.W. Slocombe was her first skipper.**

A coffer dam at Wisconsin Rapids, probably built in preparation for a new bridge on the Wisconsin River. Engineers or surveyors with their instruments appear at right center.

Above, a young crew preparing to launch on Mississippi River. Village of Alma clings to hillside in distance.

Below, fleet of lumber boats in Ashland harbor, about 1900.

Whipsaw to up-and-down saw

For thousands of years members of the human family have lived in caves, tents of pressed wool or skins, or in homes of logs, or frame buildings of whipsawed lumber, or brick, adobe, and stone. For an equal length of time the only method of sawing lumber was with a whipsaw, one man holding the handle on a long saw blade above a log while it rested on a framework, and one man in a pit holding the lower end of the saw, the so-called "pit man."

This method, still being used in China and other parts of the world, is painfully slow. But in times past there was no need to hurry. Labor was cheap. Most people, up to the time of the Industrial Revolution, worked in a semi-slave economy, beholden to families who had acquired land-wealth either through conquest, marriage, inheritance or manipulation.

In the Thirteen Colonies of America, the pace of life was also slow, and for housing, people were using stone and brick, whipsawed lumber, or lumber sawed at a locally owned sawmill with reciprocating blade. Then, suddenly, the population of the several states began to expand, and when new waves of European immigrants increased this growth, a demand arose for more building materials than farmers could supply in their locally owned sawmills.

A need arose for a new industry devoted only to the manufacture of wood products. Not only was more lumber needed, but laths, shingles, pickets for fencing, wooden tools and implements, buckets, barrels, staves and, soon, railroad ties. These demands spurred the birth of forest industries. Sawmills were built on the rapids of the big rivers to take advantage of greater power facilities which could be developed by bigger overshot or undershot water wheels. So anxious were people in the Northeast to get on with the increased pace of lumbering that bees were organized to build dams, that is, people were willing to donate their own labor to the man who put up money for a dam, But, like most things, the owner did not come away scot free. The drinks were on the house, and one dam builder in Maine, in order to keep everyone happy while they struggled with the bulkheads, ordered the following refreshments:

one bottle of brandy
one and half gallons of liquor
one barrel of cider
two bottles of liquor
three gallons of rum

But no matter how many free dams were built, demand for lumber outgrew supply. At this point, big lumber companies, backed by investment capital from Boston and New York, were incorporated to buy timber lands and to harvest the timber. In the twenty-five years from 1825 to 1850 more was learned in Maine and New Brunswick about lumbering than had been known in the past thousand years. And in the twenty-five years that followed, when lumbering spread to the Lake States of Michigan, Wisconsin, and Minnesota, production doubled and tripled. From sawing three to four thousand feet of lumber in one day on the old up-and-down sawmills, the circular saw and muley gang saws, introduced in the 1850s, increased this figure to ten thousand feet per day. The climax came in the 1880s with the perfection of the band saw when some mills in Wisconsin, like the giants at Chippewa Falls and Menomonie, running day and night, were sawing a quarter of a million feet of lumber a day.

The history of sawmilling goes back to the third or fourth century A.D., when the Romans developed some kind of reciprocating saw. The use of the waterwheel, as a source of power, was probably known by the beginning of the Christian era, and from this someone eventually made a waterwheel with an axle on it to furnish power for an irrigation lift or other machine. Practically nothing is heard about the use of the reciprocating saw again until the end of the 13th century when a French architect, Villard de Honnecourt, circa 1270, developed a crude machine for sawing logs by water power. A sketch of this machine appears to be the earliest available in Europe, although it would be difficult to tell how well it worked. But the principle of the reciprocating saw, that is, a saw that moved up and down, was now firmly established. There were sawmills in the vicinity of Augsburg (Germany) in 1337 or earlier, and literary references have been found to a sawmill on

An exhibit in the Museum of History and Technology at the Smithsonian Institution reveals how whipsawing is done, one man above the timber and one below, pushing up and down on a saw that cut only on the downward stroke. Throughout the ages, this was the only way to saw lumber before the first up-and-down saws were developed for waterpower in Europe in the 13th and 14th centuries.

the island of Madeira after 1420, in the city of Breslau, Poland, about 1427, and to two mills in a forest owned by the city of Erfurt (Germany) in 1490.

In 1558 an Italian, Agostino Ramelli, published *Dell' Artificiose Machine*, and in it appears a woodcut showing a squared timber resting on a platform, or carriage which is being impelled into a rigid saw blade suspended from a headblock in a wooden frame. The frame moved up and down *with the saw* in it and was held in place by grooves or mortises inside two fender posts, something like a window that moves up and down inside a sash. It was for this reason that these early mills were often called "sash" mills and also "gate" mills because the saw works resembled a gate, but the most common designation was "up-and-down" mill.

The sketch by Ramelli also shows four square clamps on a squared timber. These clamps acted as a vise to hold the timber rigid. They were later discarded and metal pieces or "dogs" were used to hold the square timber down.

Despite this rather advanced technique, a Virginia pioneer, Edward Williams, apparently invented an "engine," as he calls it, for sawing lumber in a mill probably inferior to the Italian mill of an earlier date. Williams had a small book published in 1650, in London, called *Virginia's Discovery of Silke Worms . . . together with the Making of the Saw-Mill*. He explains how this Rube Goldberg device worked, and also says that "this engine is very common in Norway and the mountains of Sweden, wherewith they cut great quantities of deal-bords . . . ."

It does not seem possible that any great quantity of deal-bords was ever cut with this "engine," but at least the Virginian had developed a slightly different principle for sawing logs, if not a very practical one.

The first sash, or gate, sawmills in America were built not long after the English colonists landed in Virginia. They needed lumber and planks faster than two men could whipsaw them. It is not possible, however, to determine with any degree of certainty, where the first mills were built, but tradition points to Virginia and to Newfoundland and to the years 1610 and 1620, although tradition fails to affirm which of these dates can be ascribed to which mill.

The two pictures at right and below were both taken in China in 1944. The men at the right are whipsawing in the traditional manner, up and down stroke, while the two in the lower picture are using the horizontal approach. The whipsaw is still being used in China and in parts of the world where, for local consumption, it does not always pay to install a mechanized sawmill.

In Edward Knight's great work on the history of the machine, called the *American Mechanical Dictionary* (1876), a description appears of a mill built on the Danube River in 1575. In this mill multiple "up-and-down" saws were used, later known as "gang saws." These sawed several boards at once. Similar types of gang mills were built in Holland in the late 16th century and in England in the early 17th century, but all were abandoned. Apparently the chief reason these gang mills did not change sawmilling at the time was that workmen and trade guilds were opposed to them.

Here was a new device that could eliminate a work force which, since time immemorial, had been whipsawing lumber. In several instances sawmills, and even a windmill providing power for a mill, were destroyed by mob action. (When the first mechanical steam loaders, or "jammers," were introduced in Michigan around the turn of the 20th century, they too were sabotaged by lumberjacks afraid of losing their jobs to a machine which could speed up the loading of logs.)

In the early period of sawmilling, a log was usually hewed on one side before it was placed on the carriage. The illustration from Ramelli's *Dell' Artificiose Machine* shows that the timber being sawed is already squared. This followed the practice of squaring off a log, that is, hewing it square before it was turned over to the men who were doing the whipsawing. And this practice was carried over to the sawmills. Hewed in this manner, the piece was called a "balk." In fact, there does not appear to be much difference between what Americans called a "square stick" (a squared timber) and a "balk." However, as the technique of moving a carriage forward improved, the square stick was no longer necessary, but the log was still cross-cut square at the ends and hewed lengthwise on one side to give better bearing points on the carriage.

After the log had been rolled, laboriously, it may well be imagined onto a carriage, "dogs," that is, short bars of curved iron with right angle spikes, were hammered into the log to hold it in place. By moving a lever which opened the watergates of the penstock, the head sawyer set the mill in motion. As the back end of the log neared the saw, the watergate was closed before the log reached the end to prevent the saw from buzzing the iron dogs. The carriage was then retracted to the starting position. After several boards were sawed, still clinging to the original log, or square stick, the ends were left unsawed. These unfinished ends, or "stump shots," as they were called, were then split apart with an axe or frow. It left a rough surface, but then all the surfaces left by these early saws were rough. The teeth in the saw blade of the up-and-down saw pointed down, and since the saw cut only on the down stroke, the blade left its mark on each stroke. The stroke length varied but the average ran from eighteen to twenty-four inches. In short, any log larger than two feet on the stump had to be hewed with a broad axe before it could be sawed.

Oral tradition tells of a great white pine that stood on the farm of Per Dobbe (Portage County). Jim Bigler, who operated an up-and-down saw mill on Flume Creek, offered Dobbe a sack of flour for the tree. Dobbe agreed to the trade on condition that Bigler chop the tree down and remove it. The giant was felled. The first log to be loaded on the sleigh broke one of the beams. When this was repaired, the log was finally got to the mill. Here it was found that it would require so much work to hew it down to fit the carriage that Bigler abandoned it. The two big logs that remained on the Dobbe farm were never moved. The family bucked up one to make shake shingles, and there were enough shingles to roof all the buildings and more.

The earliest visual evidence of an up-and-down sawmill in America may be this sketch reproduced life size from the *Pictorial History of the United States* published in Connecticut, 1847. Two mills may be seen at left and center, but there is no clue to the source of power that keeps these mills going this close to the sea. A ship's hull is under construction at the far right. The leisured life of a sawmill operator is reflected in the man atop the log who, knowing he has time to chat, will jump off the log before the saw gets too close to him. A chain is attached to a log below the mill to lift log into the mill, but the windlass is obscured or overlooked by the engraver. The saw machinery also seems incomplete.

Evidence of an up-and-down sawmill (right) appears in French texts about the year 1270 A.D., a saw developed by Villard de Honnecourt, an architect. It had no sash, or gate to hold the sawblade. But the blade seems to have been held by a long spring pole at the top. The long board or plank being sawed has two guide posts to hold the board in place, and some sort of a dog holding the opposite end, (left). This saw was crude but it revived the principle of the reciprocating or up-and-down saw. The illustration appears in Agostino Ramelli's *Dell' Artificiose Machine*, printed in 1588. The French text reads in translation: "With this you can make a saw cut through a beam lengthwise." The illustration below (left) appears in the same book by Ramelli, with detail of an Italian up-and-down mill of the 16th century. Below, right, an engraving of a mill developed by Edward Williams, a Virginian, in the mid-17th century.

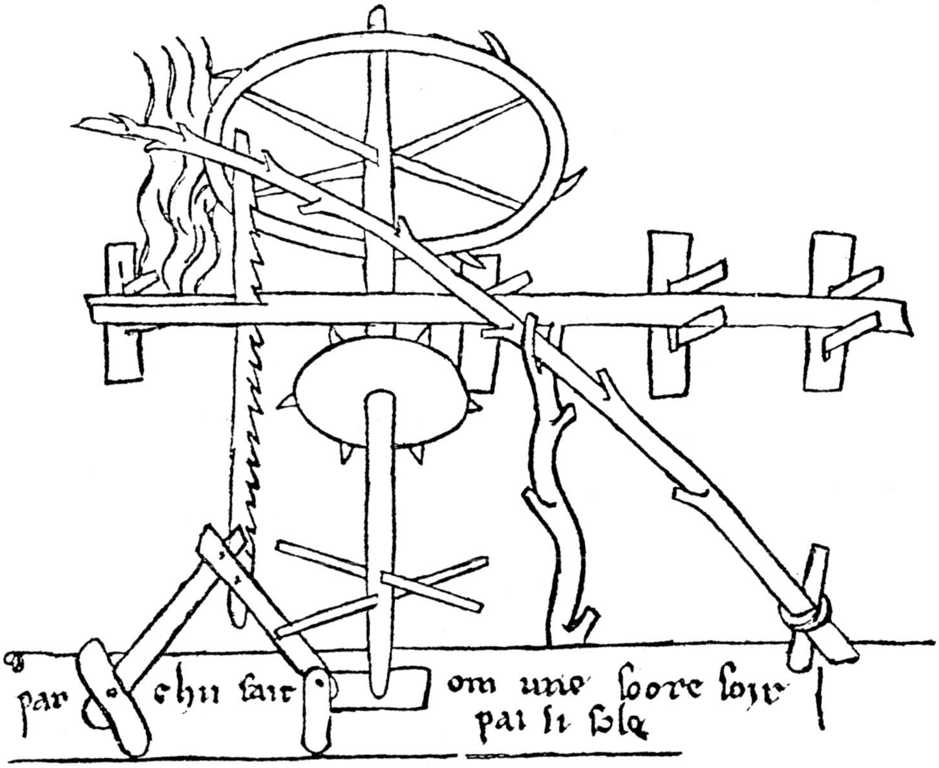

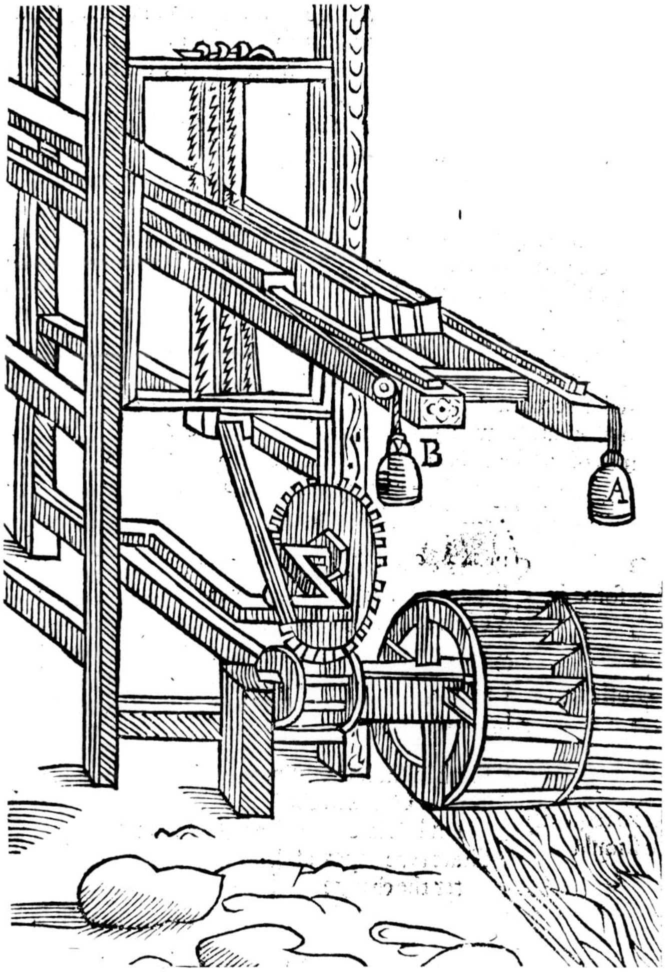

A gang mill at one of Knapp, Stout & Company's mills in Menomonie, about 1885. The logs have been squared, probably by a circular saw, and then sent to the gang saw where multiple boards will be made in one pass through the saws. Only the largest mills in Wisconsin resorted to the gang saw and most of them were discarded in favor of the band saw in the 1890s.

This, incidentally, also reflects the waste that characterized much of the logging era. It seems incredible that Bigler did not realize the limit of his own saw rig before he bartered the sack of flour, but, then, stranger things have happened. One early expedition to the South Pole forgot to bring shovels.

Another story about the Bigler mill tells of a crew that rolled a log onto the carriage, set the water wheel in motion, and then went to breakfast. By the time breakfast was finished, one board had been sawed. This is a story exaggerated for the benefit of a later generation. But the crew at a mill farther down the same stream, according to hearsay evidence, played cards while waiting for the saw to finish a board. This seems possible because a hand of cards can be laid down at any time. Both stories, true or half true, point up the unhurried, almost leisurely method of sawing logs in the small mills of Wisconsin up to 1870.

The Museum of History and Technology in Washington D.C. has on exhibit a partially restored up-and-down mill that was found in 1961 at Chester Springs, Pennsylvania. According to the descriptive plaque at the base of the exhibit, the saws of the 18th and 19th centuries in Pennsylvania made one hundred to one hundred and thirty strokes per minute. The log moved forward on the carriage at the rate of one quarter inch or more per stroke, or two feet or more per minute.

The mill at the Smithsonian Institution, described above, is probably the type that George Stevens of Almond, New York, brought with him when he came to build a mill on the Wisconsin River at "Big Bull Falls" (Wausau) in 1839. Nelson Week, who, as a boy, watched his father's mill in operation on the Big Eau Pleine River in the early 1850s, recalled that the mill sawed up to five thousand feet of lumber a day, but he failed to explain the length of the day. Many of these early up-and-down mills ran a day shift and night shift, and it is quite probable that the Week mill did the same.

The last advance in the up-and-down technique

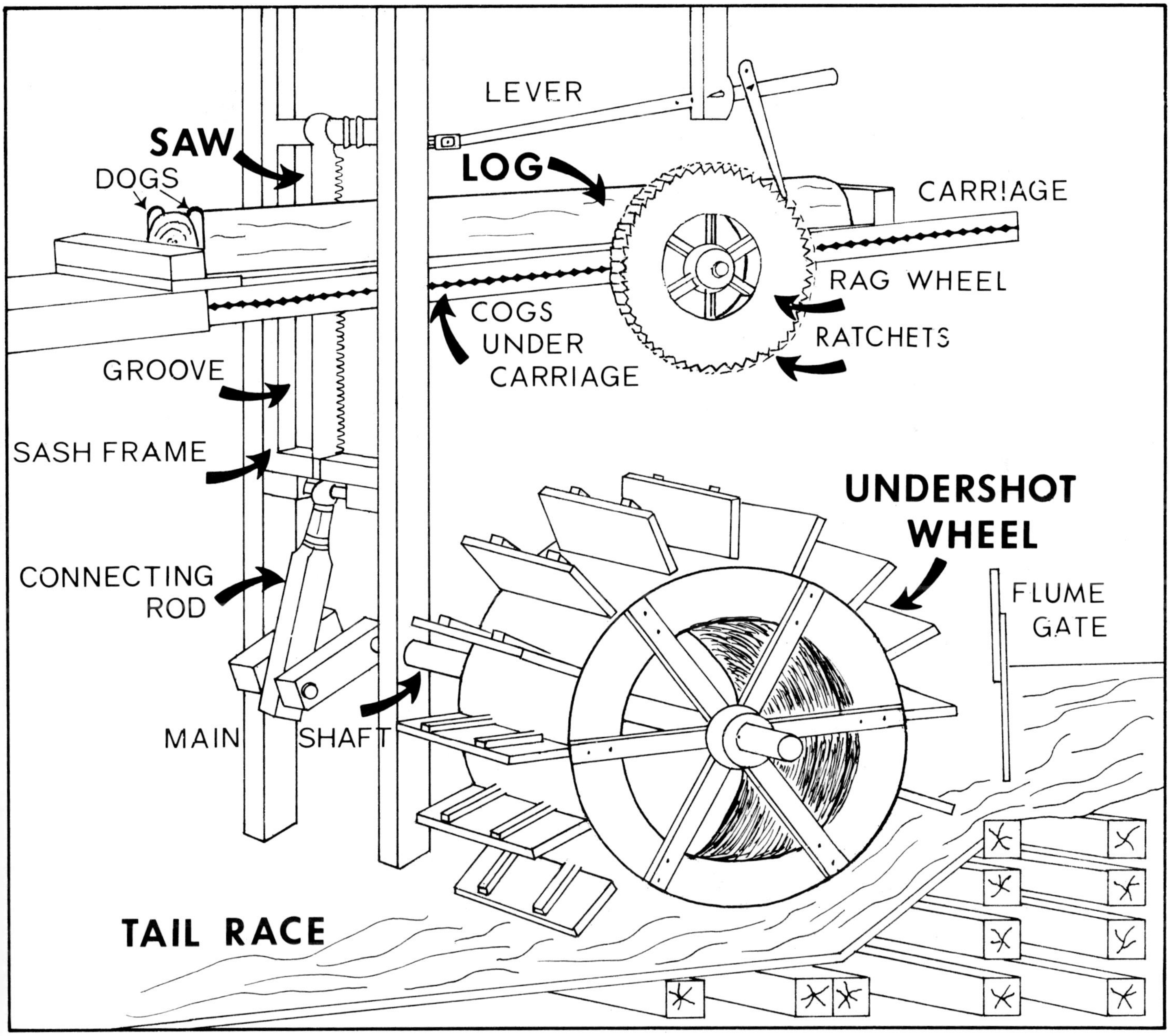

A mechanical drawing by Randall Rohe of an up-and-down sawmill powered by a waterwheel.

One of the few relics left from the days of the up-and-down saw in Wisconsin is this blade which is on exhibit at the Rhinelander Logging Museum, but the origin is uncertain. Both a saw and sash are preserved at the Dunn County Historical Society Museum at Downsville.

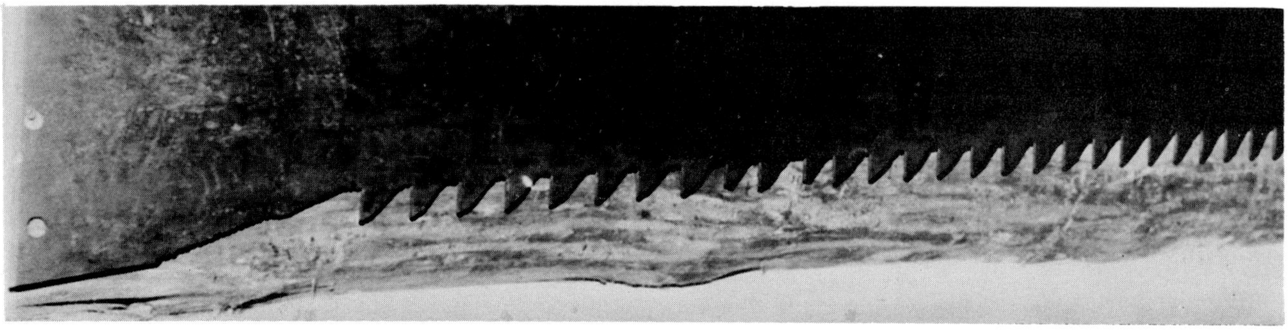

This photograph, taken at Chester Springs, Pennsylvania, before the up-and-down sawmill seen here was moved to the Museum of History and Technology in Washington D.C., in 1961, gives a fair idea of the headblock that held the suspended saw blade in place. The headblock is attached to two fender posts with grooves inside, and the whole headblock and saw moved up and down along these grooves like a window in a sash, hence the alternate name, "sash" saw, or "gate" saw.

Early types of dogs used to hold a log in place on the platform or carriage for sawing. The two photographs by the author were taken of the restored model of an up-and-down saw at the Museum of History and Technology in Washington. The position and design of the dogs explains why the up-and-down saw never sawed the entire length of the log. A life-size log has been used in this restoration, but most of these early mills sawed only logs hewed on four sides, or logs hewed on one side to enable the dogs to provide a better grip.

of sawing lumber was made with the introduction of the muley (or as mulay) saw. It was probably first used in the Thirteen Colonies in the 1780s, and by 1825, according to Barbara Benson, in a doctoral thesis on the lumber industry, the muley had come to Michigan. It was a slight improvement over the up-and-down saw because the saw blade moved up and down, not the whole frame, or sash, that held it. Although the muley, too, cut only on the down stroke, the saw was more efficient and increased production to more than six thousand feet per day in two shifts. The drawback was that it still cut a kerf nearly half an inch wide—an incredible waste—and perhaps for this reason the single muley saw was never popular with lumbermen in Wisconsin.

Meanwhile, the gang saw had come to America, and it probably was used in Michigan by the mid-1830s. Thus, when the circular saw came into general use in the late 1850s, it did not displace the sash saw mainly because of the much improved gang saws that were on the market. The main drawback to the gang saw in this period was that it could not saw a raw log. The log had to be squared off or hewed before it could be sent through the gang.

By 1874, one of Knapp, Stout & Company's sawmills at Menomonie was operating six big rotary saws and seven gang saws, and manufacturing nearly a quarter of a million feet of lumber in twenty-four hours.

The muley (or as mulay) saw that followed the up-and-down saw was not an invention of principle, nor did it replace the up-and-down saw entirely, but it speeded up production somewhat because it was more efficient in its mechanism. The saw moved up and down within its own casing; the headblock and frame remained stationary. But the muley saw did nothing to change the width of the kerf over the up-and-down saw, and it was therefore just as wasteful of good lumber as the former. This engraving appears in the *Fourteenth Annual Report of the Indiana State Board of Agriculture,* 1872.

Circular saw to bandsaw

The interior of a small sawmill in Alban township (Portage County), before 1900, operated by Ole Leklem and Dan Danielson. Captions are superimposed on the picture to show the different parts of the operation. Man standing akimbo is Peter Paulson, engineer, and at right, Osten Gutho, head sawyer who has his right hand on the lever that engaged the clutch on the carriage.

The transition from waterpower to steam was a major step forward for the up-and-down sawmills. The first to operate by steam, not by waterwheel, was probably built at Bath, Maine, in 1831. It burned slabs for fuel and was an instant success. Barbara Benson, however, believes that a Benjamin Wheelock built a steam mill in Michigan in 1825. By 1837, she says, there were twenty-two mills in the state using steam.

Thus, as the tempo of lumbering increased, more and more mills were converting to steam. One of the first in Wisconsin was probably built at Oshkosh. *A History of Northern Wisconsin* (1881) has this to say:

> Morris Firman commenced the first saw mill [steam] at this point [Oshkosh] in 1847, but Forman & Bashford of Algoma [now part of Oshkosh] completed a rival mill first and ran through the first lumber.

The Green Bay *Advocate* carried an advertisement on December 7, 1848, which had this to say:

> A steam mill just up by Mr. Hogeboom on the bank of the East River, is in operation, and works finely . . . ."

A few months later on February 9, 1849, the Oshkosh *True Democrat* reported a steam saw mill which "is now in complete operation."

Following the invention of the water turbine, patented in 1869, waterpower continued to be used more than steam by the small mills of Wisconsin. The turbine superseded the waterwheel to a large extent because it could operate in dead water as long as the "head" of water above the dam was maintained. The undershot or overshot "flutter" wheels were useless if the water below the dam backed up for any reason.

Although invented in Europe, the first American circular saw was hammered out in 1814 by Benjamin Cummings, a blacksmith in Schenectady, New York. But it was a saw in which the teeth had been filed into the outer rim of a round steel plate. Later, it was realized that the saw had to have movable teeth that could be inserted into the rim of the round steel plate. These were being advertised by the American Saw Company of New York in 1866, and portable engines for sawmills the same year. Automatic saw sharpeners for gang saws were offered for sale at least as early at June 8, 1880.

Circular saws, despite their drawbacks, were apparently being used as early as 1838. An advertisement appearing in the Bangor (Maine) *Whig and Courier* on October 9, 1838, ran this: "Wanted . . . a first rate head sawyer, two of the second class, one who understands circular saws, and one teamster . . . ."

It remains a mystery what the advertiser wanted with "two second class" sawyers, but at least, he makes it clear that he needed someone who understood circular saws. This suggests that there were still defects in it. Another two decades would pass before it would begin to replace the sash and muley saws for sawing lumber.

However, change in any industry usually lags behind technological advances, and perhaps lumbermen were the most conservative of all, for it was not until the mid-1850s before the circular saw came into general use in Wisconsin, although, as already noted, it did not displace the gang saws until many years later. One of the big reasons for the change to the circular saw was the saving in lumber. It cut a quarter-inch kerf as opposed to nearly half an inch in the sash or muley saws. Finally, it could saw one to two thousand feet of lumber in an hour.

While the band saw superseded the number of feet that could be sawed by a circular, it never displaced it. Mills in villages like Rosholt, Iola and Marion, which were doing custom sawing for farmers, never had any need for more than a circular saw, and, as time went on, a thinner saw was developed to cut a smaller kerf. It continues to be used in portable and small stationary mills.

The circular saw started out small, not for sawing lumber, but for trimming and butting lumber already sawed by the up-and-down or muley saws. As the metallurgy improved, and filing the saw became a more exact science—most mill owners paid the saw filer more than they paid the head sawyer—the saw grew in size and could cut through a log a meter in diameter. The final touch was added to circular sawing with the introduction, some time in the 1870s, of the "double saw," that is, two saws, one on the canting deck, the larger of the two, and a smaller saw, suspended over and slightly behind the main saw. This made it possible to saw almost any size log in Wisconsin.

A woodcut of a double saw machine appears in Knight's *American Mechanical Dictionary* (Vol. III) and another appears in a J.I. Case catalog for 1884 and 1885. It would be difficult to estimate how big the saws are in the Case illustration. An advertisement in the Wisconsin *Pinery* (June 7, 1855) reveals that at this early date, saws six feet and six inches in diameter were on the market.

While the circular saw was a big improvement over the reciprocating saw of the early 19th century, some historians believe that the greatest boost in forest industries was made by the invention of the planing machine, and nail making machine. Carpenters were relieved of the time-consuming task of hand-planing the rough lumber, and machine-made nails eliminated the cumbersome "square" nails made by the local blacksmith, and could be manufactured cheaply and in quantity.

Samuel Bentham, an English engineer, inventor and historian, is generally credited with the invention of the first practical planing machine for wood, circa 1791. By the year 1820 specifications for nine planing machines were allegedly patented in the United States, the most revolutionary one taken out by William Woodworth of New York state who had found a way to use feed rollers together with the rotary cutters that planed the rough surface of the board. This great labor-saving device, too, was opposed by carpenters and joiners, fearful of competition with

Below are two illustrations from the J.I. Case Company catalog of 1884, the one at the top showing a portable sawmill powered by a steam engine resembling the engine used on a Case threshing machine. A double saw is in operation at the left. The logs are drawn out of proportion to the tools and to the men handling them. This was summer logging and highwheel carts are being used for skidding. The stack on the engine has no hood and the forest is in imminent danger of catching on fire.

The illustration in the lower half of the page, also from the Case catalog of 1884—and 1885— provides one of the first glimpses of the carriage on which the logs rested while being sawed, sometimes referred to as the headrig. The dogs, or hooks used to secure the log against the headbocks on the carriage, are also among the first ever to be illustrated in this manner. This model was later replaced by more efficient ones in the next decade or so. The inset at the left is a "set works." It can be seen on the log shaft at the back of the carriage between the two headblocks. The lever was used to nudge the log forward an inch at a time. The inset at the right illustrates belt tighteners for sale by the company.

The Paine Lumber Company of Oshkosh used this illustration in a commemorative booklet. The inscription at the left was apparently inscribed by the advertising department over the original illustration used in a catalog of the 1870s. The illustration confirms that bandsaws were in use in some mills, though most mill owners were holding off, and it was not until the 1880s that bandsaws, like the one shown here, were installed in the larger mills of the north. This illustration also provides a closeup of a carriage in use in the 1870s, and the set works, one behind each headblock. The dogs also seem more efficient than the ones shown in the J.I. Case catalog for 1884. But both Case and Paine were using a carriage that operated on iron wheels running over some sort of strap iron mounted on wooden rails. The door in the illustration was part of an advertisement for the Paine company which was manufacturing doors, windows, sash, interior trim, etc., not lumber per se.

time-honored prerogatives. The Woodworth patent, which monopolized the market for many years, was finally broken in 1865, and better and faster planing machines were developed.

The introduction of the saw carriage on wheels, with strap iron laid over wooden rails for tracks, probably came in with the circular saw in the early 1850s. And with that also came the modern sawmill, in a sense, because it was not operated on cogs and ratchets, but by belts and pulleys off a main shaft powered by a steam engine or turbine.

The iron dogs shown in the Case illustration have changed since the 1884 catalog. Instead of the clumsy clamps shown in the catalog, the new dogs are attached to the front of the headblock, and when a log is rolled onto the carriage, the dogs are dropped by the carriage rider to secure them, one dog to each headblock, usually three in number. A simple lever on the dog disengaged it when the log had to be turned or removed. Knight carries an engraving of a carriage with headblock and dogs as they were used prior to 1876.

The illustration in the Case catalog probably provides the first evidence of the type of dogs that

By the turn of the 20th century, bandsaws of the size shown here, were being manufactured in Eau Claire by the McDonough Manufacturing Company, a company founded in 1888 by Frank McDonough, a pioneer tool maker. No carriage was used for the bandsaw; two-inch planks, apparently, were run through the rollers for resawing.

were being used on the headrig to clamp down and hold the log secure. The leverage setworks (see in the inset) combined the principle of rack and pinion, and moved the log forward on a graduated scale after each run through the saw. The sawn lumber piece dropped off the log automatically and was picked up by a man called the "edger" who sent the piece down a long table with rollers, to edging saws for trimming.

It was the duty of the "setter" to handle the set-works on the carriage. A second man on the carriage, the "rider," handled the dogs, although the setter helped with the dogs too, and if the crew was short handed, the setter took care of the entire works. It kept him jumping.

There is little similarity between the dogs seen in the Knight-Case illustrations and the several newer types developed later and reproduced in

Fig. 4629.

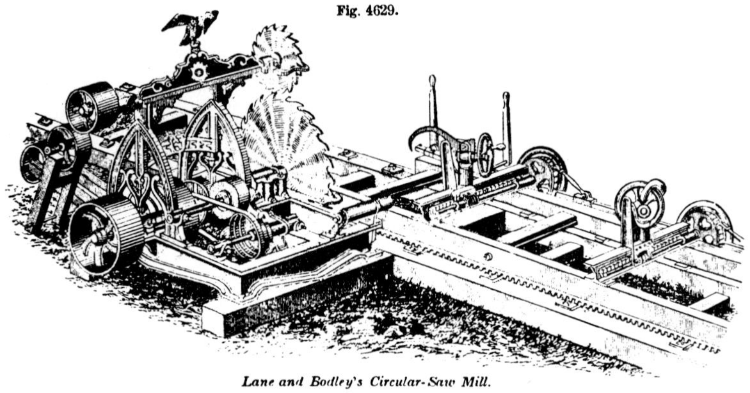

Lane and Bodley's Circular-Saw Mill.

This illustration, actual size, appears in the *American Mechanical Dictionary* (1876), and it shows a double-saw, and also another type of carriage. What is unusual about it is the type of dogs used to hold the logs in place on the carriage. The dog mechanism apparently served also as headblock. In an illustration of "Lane and Bodley's Circular Saw Mill" is a sketch of a headblock which does not resemble the Lane-Bodley patent, and which was probably more efficient. At the left is a figure of a "saw-mill dog," which seems to be more primitive than either of the former.

adopted because circular saws or disks have been used for many centuries in lathes and lapidary work. Miller's English patent of 1777 describes the circular saw, and such were made at the commencement of the century by General Sir Samuel Bentham for the British Admiralty. At present the circular-saw mill is the favorite in the woods, where the location is to be occasionally changed to comport with the supply of logs. Circular saws of very large diameter are also used in some of the great mills on the Mississippi. See also CIRCULAR SAW.

Saw-mill Dog. A device for holding logs on the carriage while being sawed.

Fig. 4629 has a swivel joint to permit the point to be turned in any direction.

Fig. 4629.

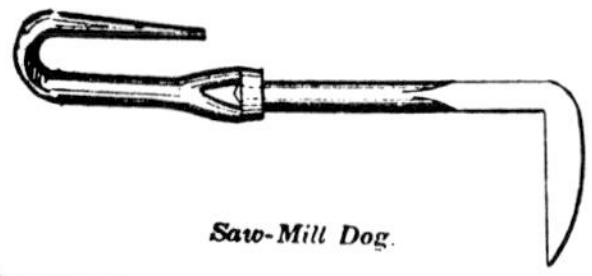

Saw-Mill Dog.

Fig. 4630 has an arrangement of two dogs that hold the log while it is being sawed, and are adjustable by means of a screw passing through the dog-heads. Two parallel rods also pass through the dog-heads to keep them firmly in position, and all turn or swing upon an upright rod passing through a swinging head, and are again adjusted by a screw-rod passing through the swinging head and bearing on the top of the head-block.

Saw-mill Gate. See SAW-GATE.

Saw-pad. A contrivance for conducting the web of a compass-saw or lock-saw in cutting out small holes.

Fig. 4630.

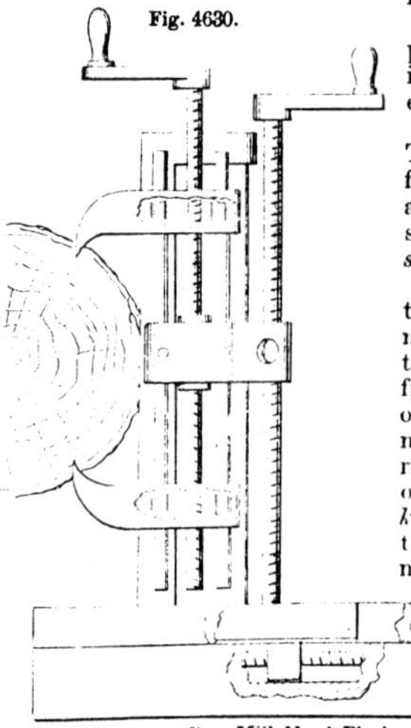

Saw-Mill Head-Block.

Saw-pit. The pit beneath a log in which the lower sawyer works.

Saw-sash. The rectangular frame in which a mill-saw is stretched. A *saw-gate*.

Saw-set. A tool or implement to slant the teeth laterally from the plane of the saw, alternately to the right and left, in order that the *kerf* may be wider than the thickness of the blade, and friction be reduced. In some cases, the edge of the tooth is spread to widen its cut, instead of bending it laterally.

mechanics drawings appearing in C.J. Telford's *Small Sawmill Operator's Manual* (Agriculture Handbook, No. 27, 1952).

The improvement in the quality of the circular saw, and its size, over the three decades after it was introduced in Wisconsin in the early 1850s, is reflected in production figures at Bayfield in the "Little Daisy" mill founded by Captain Robinson S. Pike in 1870. In 1888, this mill, with one big circular saw, was sawing seven thousand feet per hour, six days a week on an eleven hour day. By 1893 the mill was sawing two hundred and fifty thousand a day, although the statistics fail to explain whether this was with one saw or a double saw. If it was with one saw, much of the sawing may have been of two-inch stuff which was later passed through a re-saw to make the standard one-inch stuff.

A correspondent from Chippewa Falls, in a letter to the Wisconsin *Pinery* on April 26, 1865, had this to report on the subject of saws:

> Messrs. H.S. Allen & Co., have extensive saw mills on these falls, running five upright [saws] and two circular saws, besides a gang of fifteen more [saws]. Their daily product of lumber is 60,000 feet.

This same correspondent ended his report to the *Pinery* on an optimistic note: "The pine lumber appears to be inexhaustible above [the falls] . . . ."

One of the big problems faced by the early circular sawmill operators was how to "cant" or roll a log onto the carriage. For small logs this was no problem. It was customary for the "canter" to roll a log onto the headblocks of the carriage, and, using the helv of the cant hook as a lever, he moved the log up tight against the headblocks while waiting for the carriage rider to drop the dogs on it. But many of the logs that came into the larger mills were too big for one man to handle. It is said that in some of the early mills a block and tackle, suspended from the ceiling of the mill over the carriage, was used to skid the large logs up to the headblocks. When the dogs were secured, the tackle was lifted away for the pass through the saw. This must have been time-consuming, but no picture of this device seems available.

Finally, to make it easier to mount a big log on the headblocks, Alexander Rodger of Muskegon, Michigan, developed the "Tarrant Log Turner," also known as the "friction-nigger." A double bar friction log turner, and a single-bar friction log turner are both illustrated in the *Small Sawmill Operator's Manual.* But the type of log turner used in the majority of Wisconsin sawmills was the rocker which is not illustrated in the manual. This had wooden sockets at the ends of a long bar, and a third socket in the center. When a log was rolled up to the carriage by the canter-on-deck, the rocker shot up from a well deck and kicked the log firmly against the headblocks until the carriage rider lowered the dogs on it. After the first slab on the log had been sawed off, the carriage moved back to its starting position. The dogs were yanked up and the log turner shot up from the deck again to juggle and turn the log around.

A story appearing in the Wisconsin *Lumberman* on February 3, 1864, reveals that William Avery of Stevens Point had installed the first device "for canting logs athwart the headblocks on the sawmill carriage."

But Avery had not installed the Tarrant Log Turner invented by Rodger. An advertisement in the *Pinery* on November 3, 1865, reveals that he was using "Foster's Patent Device for Canting and Turning down logs . . . ." John Week on the Big Eau Pleine and John C. Clarke in Wausau had also installed the Foster patent. Thus, within a short time there were already two log turners to choose from.

Smaller mills got along without the new device, leaving it up to the canter-on-deck to turn the log. Most mills, after the turn of the 20th century, were sawing logs averaging less than half a meter on the stump. Farmers who brought logs into the mill for custom sawing were often cutting what early-day lumbermen had ignored, particularly hemlock and hardwoods. A veteran of the logging woods, the late Ray Brunuick, once told Harry Curran (Rib Lake) that "we wouldn't look at hardwood until 1900. Hemlock in the 1880s was worthless as timber. The bark was peeled off and the logs left there . . . ."

Installation of the bandsaw in the larger mills was the final step in the revolution that had been taking place in sawmilling since the beginning of the 19th century. Bandsaws, it seemed, could saw more logs than the men in the woods could cut down, and that required bigger crews, bigger camps, and even bigger horses.

Although invented in England and developed in the United States in 1817, the bandsaw remained an imperfect instrument, lacking the right quality of steel to withstand the wear and stress of sawing the big American white pines. An "improved bandsaw" was exhibited at the Philadelphia Centennial in 1876, although another was being advertised in the Northwest *Lumberman* in 1874. In 1876, the J.I. Case Company of Racine began manufacturing the first line of bandsaws in the state, but there were still "bugs" in it, and mill owners held back. Finally, by the mid-1880s, the bandsaw was given the seal of approval by the lumbermen. It not only cut a log much faster than a rotary saw, but it left a smaller kerf and saved millions of board feet of lumber that would otherwise have gone into the slab burner. It was a triumph of metallurgy and engineering. Ironically, it was also a disaster. The big pines that many lumbermen assured themselves would last forever, were running out. Nevertheless, by the 1890s many of the larger mills in Wisconsin were installing the bandsaw. (One of the best examples of the modern bandsaw may be seen at the sawmill of Menominee Tribal Enterprises at Neopit.)

But, while all these innovations and inventions were replacing older methods of sawmilling in the United States, lumbermen in Canada were still not

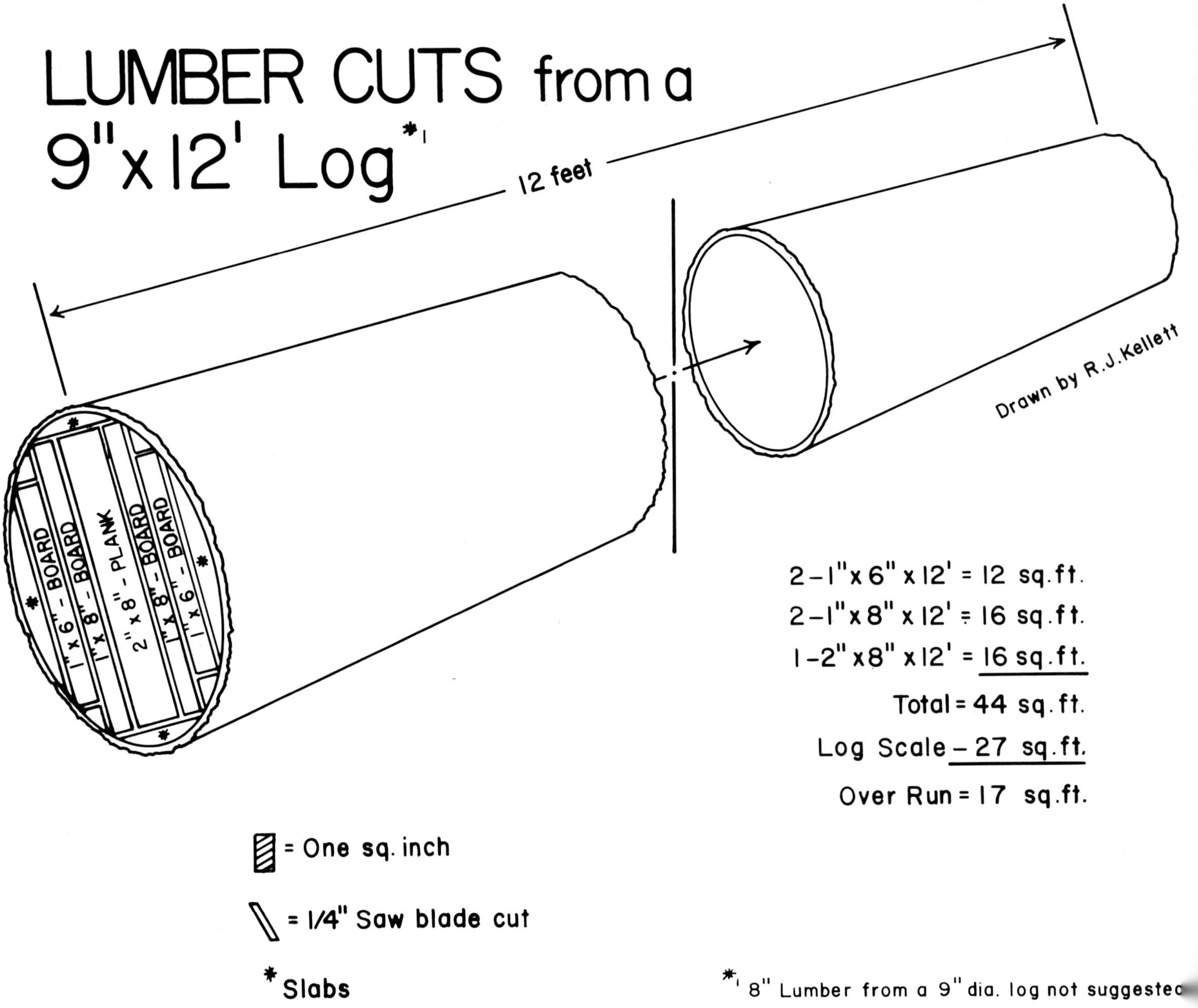

If a log were perfectly round at both ends, this is how it would be sawed into boards, according to an estimate made by Lester Peterson (Rosholt).

convinced that the rotary saw was ready to replace the gang saw. According to Knight, one big lumber company in Canada, Perley & Pattee, was sawing 250,000 to 300,000 feet of lumber every twenty-four hours, and the owners were doing it with slabbing gang saws, stock gang saws, and Yankee gang saws, succeeded by double-edgers and double-butters.

There was no carriage in this Canadian mill. After the log was placed on the skidway or deck in the mill, it was held in place by dogs on an endless chain which advanced the log first to the slabbing gang saw and from there on to the other gang saws. Without photographs, it remains difficult to visualize this Canadian mill. The slabbing saws described by Knight were, he says, five to seven feet long, the stock saws four and a half feet long. Thus, an entire mill was being operated with reciprocating saws not unlike the sash saws of a much earlier period, the big difference being that these saws cut on the *up* as well as the *down* stroke.

It would appear that no circular saws were being used by Perley & Pattee, at least prior to 1876, although the circular saw by now was being widely used in the United States. The Basin Mill at Orone, Maine, according to Knight, used four gang saws, five single saws and two circular saws. He adds that "the circular sawmill is the favorite in the woods, where the location is to be occasionally changed to comport with the supply of logs." Here he was no doubt thinking of the portable sawmill, apparently unaware of the large permanent installations in Michigan and Wisconsin which were using big rotary saws and cutting probably as much as Perley & Pattee and wasting much less lumber.

In the United States, meanwhile, it was customary in the 1850s and 1860s to "slab," that is, saw a log on four sides with an up-and-down saw to get the bark off first. The "cant" or square stick (in Canada and England called the "balk") thus produced, was then mounted on the carriage and sawed into boards with the rotary saw. Perhaps this was done to save the teeth on the rotary saw. Logs were not always soaked in a mill pond or river before being sawed, and the bark was apt to be infiltered with sand and debris. (At Neopit a "debarker" removes the entire bark before the log is sent to the big bandsaw.)

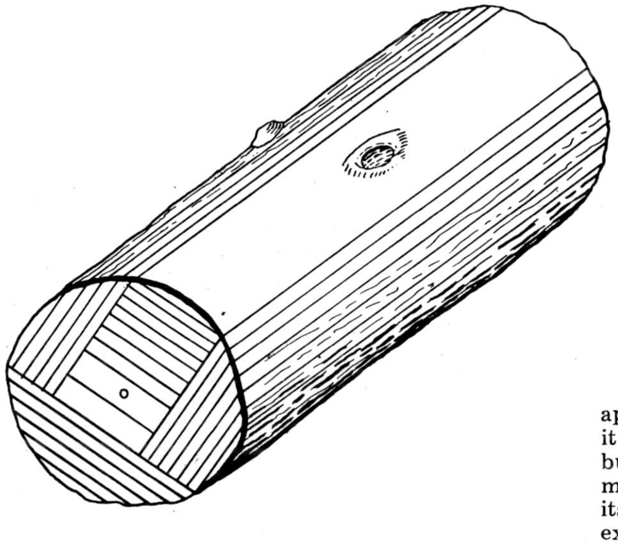

In *The Small Sawmill Operator's Manual* (1952) appears a diagram which explains how a log should be cut if it has a small sound knot and a large dead knot. It was the business of the head sawyer to estimate how to get the most out of a log, not only by the knots or rot in it, but by its curves, and by the variation in size of the two ends. For example, most logs are larger at the butt, and the head sawyer had to signal to the carriage riders how he wanted them to move the setworks forward to get an inch or two inch piece of lumber.

The bull slide

In times past, logs were skidded into the sawmill by oxen, often called "bulls," over a ramp that ascended only a few feet, since the sawmill deck was still on the ground floor of these early mills. This may be how the ramp got its name—the "bull slide," from the bulls that used it. Later on some mills used a long rope or chain on a windlass to skid the logs up the bull slide into the mill. A picture taken of a sawmill in Rosholt in the late 1880s shows a bull slide coming directly up from the mill pond into the mill. The grade is not steep and either oxen or horses could have been used to skid the logs into the mill. This mill burned in 1901, and it was replaced in 1902 by a two-story mill, but in addition to a bull slide, an endless chain was installed. This chain, commonly called the "bull chain," went up, over and under the bull slide like an escalator in a modern department store, traveling under the water in its downward thrust to hook a log which was pushed against it by workers on the pond and carrying the log up to the second floor of the mill.

Another improvement in sawmilling came with the introduction of the "hot pond." This was a body of water set off by rough planking at the base of the sawmill jutting into the local mill pond. The icy water was piped into the boiler room of the mill where it was heated and then piped back. The hot pond made it possible for the mill to operate all winter. The logs were soaked in the warm water, and ice and debris melted away from it. The Northern Wisconsin Lumber Company at Ashland and the Williams & Salsich mill at Star Lake both had hot ponds about the year 1900.

Slabs from the first cut on a log were used to heat the boilers, but the larger slabs were also used in the manufacture of laths. Most frame houses of the 19th century were using lath over the studdings to be covered with plaster. As a result there was always a big demand for laths.

In addition to the manufacture of laths, one of the biggest subsidiary industries to grow out of lumbering was the manufacture of shingles. Up to the 1850s, most shingles were made from clear pine, usually called "cork" or "pumpkin" pine. This was not a new variety of pine, but it was supposed to be pine entirely clear of knots. There were lumber companies who advertised themselves in the American *Lumberman* around the turn of the 20th century as dealers in "cork pine" only.

To make shingles by hand, a pine log was sawed into "bolts" a meter long or so. An L-shaped frow was used to split the bolt. After a shingle had been split off, it was placed on a carpenter's bench (in German called the *schnitzelbank*), and with the operator's foot on a pedal underneath the bench, the shingle was held in place while the operator used a draw shave to whittle down one end of the shingle. For this reason, the shingle, originally called a "shake," was also known as a "shave shingle."

An advertisement in the *Pinery* on April 3, 1865, describes a sawmill in Oshkosh which had installed "Morrison's Patent shingle machine [which] could rive, shave, and joint 50,000 shingles per day . . . ." Although hand-made shingles continued to be made among farmers for their own use, the commercial shingle was now entirely machine-made. Robert Law advertised in the Wood County *Republican* (January 5, 1858) that he had a "new invention," a shingle machine which had been installed in Robins Mill (at Wisconsin Rapids), apparently unwilling to recognize earlier shingle machine patents.

Adolph Kenneberg, mentioned earlier regarding a survey made for the Wisconsin Public Service Commission in 1944, writes that the first sawmill in the future state of Wisconsin was built at DePere in 1809. The first mill in Wisconsin of which there is solid evidence was built near Green Bay in 1826 on land leased from the Menominee Indians to John B. Arndt. Other leases followed in the next several years to James Duane Doty, Daniel Whitney, Amble Grignon II, et al. The lumbermen came not only to Green Bay, but spread north from Prairie du Chien to the Black River country, and up the Chippewa River, and the upper Wisconsin River where, as mentioned before, Daniel Whitney built a mill at the first rapids (opposite the present city of Nekoosa) probably in 1832.

Colonel John Shaw told Lyman Draper in an interview given in 1855 that he built a mill at the "first fall" on Black River in 1819. This mill site

The Salsich & Wilson Lumber Company at Star Lake, about the year 1899, installed a "bull-slide" and "bull-chain" which was used to convey logs from the mill pond to the second deck of the mill. Two "pond monkeys" are seen here shoving the logs onto the chain, using pike poles. The big stack on the right is the slab and sawdust burner.

was probably near the present city of Black River Falls. There is no hard evidence, however, that Shaw built this mill in 1819, and reports of other mills in Wisconsin before 1826 cannot be verified, including the one at DePere.

In the next fifty years, there was scarcely a sawmill in Wisconsin that did not burn down. Most were rebuilt, and often after a few years, burned again. J.G. Rosholt (Portage County) never carried any insurance on his mill or lumber yard. In a fire that destroyed both in 1901, he lost a million feet of lumber, and the mill, but he said he came out even. If he had been paying insurance premiums since 1885, when the mill went into operation, he would have spent $12,000 which is all the mill and the lumber were worth.

Another angle on this comes from the Oconto *Pioneer,* which on July 17, 1862, had this to report:

> The new shingle mill of Mr. Henry Nason at Menominee was burned on the morning of the 15th inst. It was a new mill, and had only been running two weeks; cost $4,500, and was a total loss. Mr. Nason is pecularly unfortunate in his steam mills. But a little over one year ago, he had his mill totally destroyed by the ice from the bay [Green Bay] jamming against it . . . by a northeast wind . . . .

By 1894, Wisconsin had become the leading producer of forest products in the nation, and there was scarcely a city or village in northern Wisconsin without its own sawmill or shingle mill. In one city alone, such as Marinette, according to Randall Rohe, there were nine big sawmills in

GROUND PLAN OF A HOT POND

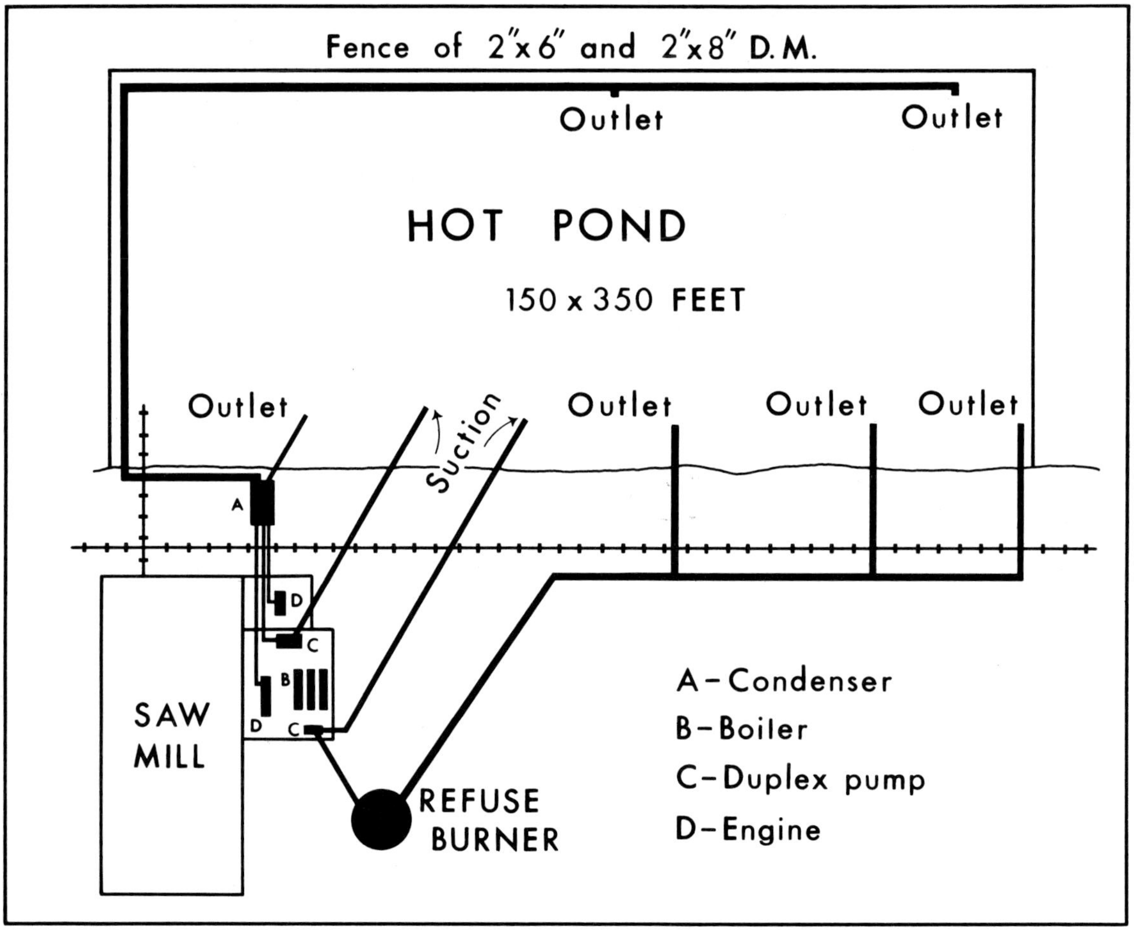

1895, namely, the Stephenson Manufacturing Company, N. Ludington Company, Hamilton & Merryman Company, H. Witbeck Company, Marinette Saw Mill, R.W. Merryman Company, Sawyer-Goodman Company, Menominee River Lumber Company, and Marinette Lumber Company. These mills were all draining the pine from the great Menominee River basin extending along the Wisconsin-Michigan border.

Down to 1898, it was still possible for mill owners to dump their sawdust or any other waste into the rivers. By an act of that year, entitled "Of Offenses Against Public Policy," this became illegal. In many small villages, the local mill pond or river continued to be used to dispose of garbage or old mattresses, and creameries continued to pollute local streams with whey before they were stopped by action of the Department of Natural Resources.

After 1915, the Water Power Law conferred upon the Public Service Commission the power to hear and determine matters respecting permits for new dams in navigable streams, power to regulate the flow of water, and the duty to report to the Governor any obstructions to navigable waters. This authority has since been absorbed by the DNR.

Modern man may be tempted to look back on the slow development of the sawmill since the 13th century and wonder how anything ever got done in the building trades, or how people managed to get anything built of wood before there were better sawmills to meet the needs of the architects and engineers of the Middle Ages. But human patience and skill have known no frontier, time or age. Consider, for instance, the carefully crafted furniture found in the tomb of Tutankhamen. Consider too the great cathedrals of France and England dating to the 12th and 13th centuries. Consider the scaffolding that was used to move the famous Egyptian obelisk from Nero's circus to St. Peter's Square in Rome, in 1586, so graphically illustrated in Bern Dibner's *Moving the Obelisks.* (This illustration was presumably engraved two years before Ramelli published his *Dell' Artificiose Machine.*)

Interestingly, the illustration in Dibner which

The double-friction log turner (above), illustrated in the *Small Sawmill Operator's Manual,* is probably similar to the type developed by Alexander Rodgers of Muskegon, Michigan, about 1860. By 1899 a steam log turner was being advertised in the American *Lumberman* which did not have a "tooth bar" like the one shown here, and which was manufactured in Oshkosh under the trade name of "The Simonson." Below, interior of a mill at Newald (Forest County) which shows three wooden sockets holding back a small birch log in center. This was the rocker arm type of log turner, probably the most widely used in Wisconsin after the turn of the 20th century.

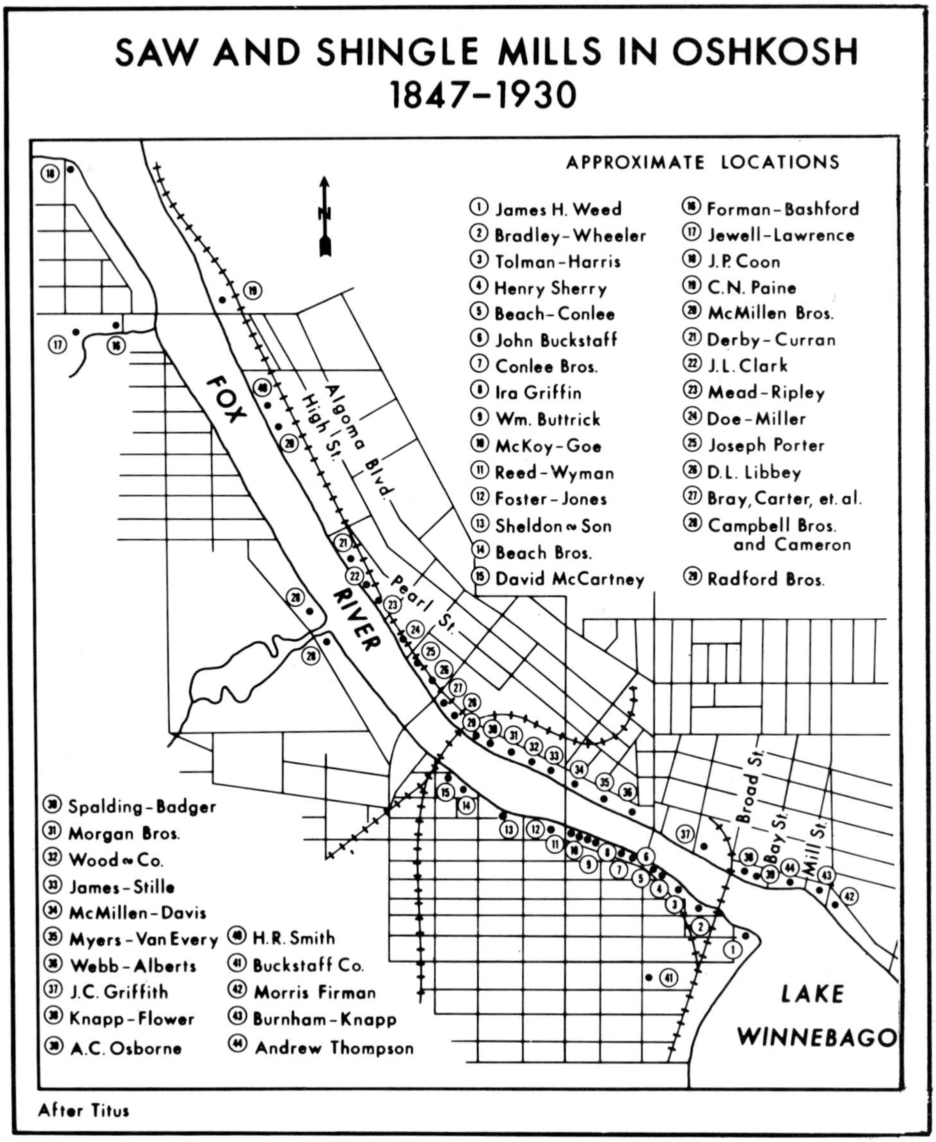

A diagram by Randall Rohe gives the name and location of the many sawmills in Oshkosh, once called "Sawdust City."

shows the scaffolding used to move the obelisk, also shows the use of the four-wheel sweeps and ropes running from the sweeps to the toe blocks and upward through the works to the obelisk—the same type of sweeps that were used by Jack Hunter in the 1870s to remove the rocks and windfalls from the Little Wolf River!

Since 1940, logging has become largely mechanized, and there appears to be no end to the number of ways left to remove a tree from its stump and transport it out of the woods. But probably the single most useful machine for mankind as a whole is the chain or "power" saw. The idea that a motor-driven saw could be used to fell a tree, or buck it up, no doubt had occurred to a number of machinists and mechanics. Walter Hall, author of *Barnacle Parp's Chain Saw Guide* (1977), credits Andreas Stihl of Germany with the final invention of the chain saw but he refuses to say whether it was invented "in 1926 or 1927, depending how you look at it." Actually, Stihl invented two machines that resembled the chain saw of today.

The first chain saws were unwieldly machines which took two men to handle, almost like a cross cut saw. But the principle of using a chain on a bar caught on quickly and American companies were soon in the market with a similar saw in the 1930s. By the 1950s the small, single-handed, sixteen to twenty-inch bar was being produced by the thousands and sold around the world. Now, even the trees in the jungles of Brazil and Borneo are no longer too high or too thick to cut down. All the trees in the world are at the mercy of the chain saw.

The sign above the overhang of the building at the left reads: "I. Stephenson & Co." The building was probably the office and living quarters of Isaac Stephenson. In his biography, *Recollections of a Long Life,* Stephenson says he took possession of the mill on Mission Point (right) on the Menominee River at Marinette in 1858. The mill was built by N. Ludington who ran it only two months and closed down. Stephenson reopened the mill and this marked the beginning of a career that ended in the United States Senate.

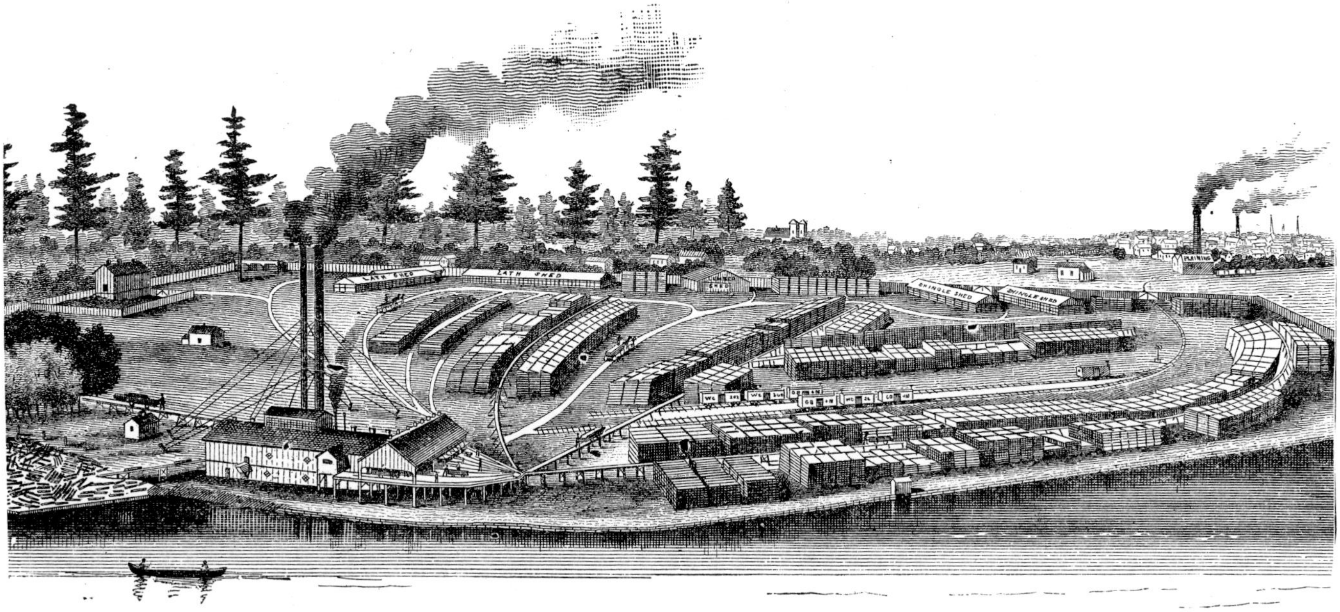

Although there were great photographers using cumbersome cameras in the 19th century, there was no way that their pictures could be reproduced in newsprint until March 4, 1880. The best that printers could do was to make an engraving from a photograph or tintype. The Bosworth & Reilly mill at Stevens Point, shown here, stood on the east bank of the Wisconsin River at the west end of present Washington Street. The view looks east toward downtown Stevens Point, which seems a long way off, but is actually only a few blocks. This engraving, lifted from *A History of Northern Wisconsin* (1881), shows no bull-slide running into the river. The log pond appears to be at the left, and the buildings at left and right are shingle and lath sheds, and at far right, the planing mill.

Sawmill of Vilas County Lumber Company, one of largest in Wisconsin, was located at Winegar on Presque Isle River. Community was first known as Fosterville, after J.J. Foster, walking boss for the company when mill was built in 1905. William S. Winegar of Michigan acquired mill in 1910 and village became known as "Winegar." After a few years mill was acquired by William Boniface Lumber Company, which operated into 1920s and closed down. Machinery and lumber was sold for salvage during World War II. At height of operations after 1910, 300 men and women worked in sawmill, yards, shops, boarding houses, company store, and in logging woods. Lumber was shipped out on Chicago & North Western branch to Mercer.

The big sawmill of the Rust-Owen Lumber Company at Drummond (Bayfield County) in 1911. At the right and left stand lumber yards, and in the distance the main line of the Chicago, St. Paul, Minneapolis & Omaha railroad. In this picture logs on flat cars are ready to be unloaded into the hot pond in the foreground and from there they will be sent up the "bull-slide" to the second deck of the mill. The slab burner stands at the left; a conveyor carried slabs and sawdust into the burner. The first addition of the mill was built in 1882-83, and it closed November 7, 1930. The officers of incorporation in 1882 were John S. Owen, president, F.H. Drummond, vice president, and R.E. Rust, secretary-treasurer. The company owned 80,000 acres of timber which was purchased from the Omaha railroad and in the nearly fifty years of operations, more than a billion board feet of lumber was sawed.

The Ehlinger double sawmill at Hollister on Nine Mile Creek in Langlade County, about 1925. This mill was one of the last to operate in the Wolf River Valley. The crew at this time included, left, Nick Ehlinger, head sawyer, Herman Miller, Frank Allen, John Mathys, Irvin Stutzman, Richard Rose, and Michael Ehlinger Sr., who served as saw filer. In back, with hand on Rose's shoulder, Thomas Rigney.

In top photograph an engine of the Wisconsin, Ruby & Southern railway (listed in the 1911 *Gazetteer*), which connected with the Chicago, St. Paul, Minneapolis & Omaha line. The village once had a population of one hundred and twenty people, most of whom were employed by the Ruby Lumber Company (lower photo), located about fifty kilometers northeast of Chippewa Falls. E.L. Haun, founder of the firm of Sill & Haun, came to this area in 1902 and founded the Ruby Lumber Company, named after his only daughter.

The T.B. Scott Lumber Company at Merrill, about 1900. The mill, largest in Lincoln County, was located on the Wisconsin River at the east end of Mill Street. The silhouette of the former city hall tower may be seen in the distance at right.

This picture from the 1880s was probably taken of a lumber mill crew at La Crosse, and includes a number of women who may have been employed as shingle packers.

The John Paul Lumber Company mill at La Crosse, about 1892. In the foreground, the frozen Black River. This was one of the few mills in the state that used a hood in the stacks to guard against flying embers.

The Crocker Chair Company had its sawmill in Antigo, and camps to the east. This wintry scene was probably taken in the 1890s. The sawmill has a "bull-slide," but no "bull-chain." The logs were being skidded by horses up the ramp into the mill. Note the height of the skidways at right.

The sign on the ridge board (upper left) reads: "Northwestern Lumber Company." The Northwestern organization controlled seven sawmills at one time, two in Eau Claire, two in Potterville, one each at Stanley, Gile, and Sterling. The mill shown here was probably located on the Eau Claire River near its mouth on the Chippewa. The picture was taken before 1907, the year the company dissolved. The property was originally purchased from Mississippi River Logging Company, and by 1882 was sawing thirty-eight million feet of lumber annually, presumably at Eau Claire. In 1867 there were twenty-two saw and/or shingle mills in Eau Claire, reaching all the way through the city and north on the Chippewa River to Blue Hill Pond, above the Dells. In 1877 there were only seven left. The giants had swallowed the small mills unable to remain solvent when losses occurred. In this picture note the heavily loaded "dump wagon" crossing the bridge, no doubt slab wood for local kitchens.

The Knoke Lumber Company sawmill crew at Hatley (Marathon County), 1913. Otto Knoke, the owner, sits fourth from the right, hands clasping right knee. Elmer, a son, suspenders showing, stands behind man in center. Standing behind his father is Armin, and a third son, Melvin, peeks over the head of the man at extreme right. Knoke began sawing in Birnamwood in 1906, moved to Hatley in 1913, and to Appleton in 1920.

One of the owners of the Nash Lumber Company of Wood County may have been from County Shanagolden, Ireland, for when the company built this sawmill west of Glidden on the East Fork of the Chippewa River, the community that grew around the mill was called "Shanagolden." The footings of the old boiler room are still visible; the rest of the village has vanished except for the town hall.

W.O. Baum, one of the officers of the Bowler Lumber Company, had a camera and a tripod with a revolving axis that could take panoramic views, like this one. The skidways of logs in center have been decked with the help of the swing-boom jammer seen at the left. The sawmill stood on the site of the present ball park in Bowler (Shawano County).

The mill at Stanley, operated by the Northwestern Lumber Company at Eau Claire, in which John S. Owen was the principal stockholder , and who gave his name to the village of Owen.

The crew of the Alexander Stewart Lumber Company at Wausau, probably late 1880s. The mill stood on the same site as the former mill of Walter McIndoe, pioneer to Wausau in 1847. McIndoe bought George Stevens' mill built in 1839-40, but built a new mill a short distance north of the one-time Stevens Mill at the foot of Forest Street. In the picture below, the company boarding house stands on the hill at the right, a site currently associated with the south end of 2nd Street. Across the yard from the boarding house stands the company horse barn. Arthur Hase (Wausau), who has preserved this picture, notes that none of the crew is wearing glasses.

The Medford Lumber Company mill appears to have been built on or near the site of the former McCartney-Whelen mill (see below).

One of the first sawmills in Medford, this one was operated for a time by David McCartney and T.C. Whelen. Apparently McCartney felt so strongly about his family he had their pictures inset in this postcard which was distributed for publicity purposes. McCartney emigrated, with his parents, from Ireland to Ohio. Later, he moved to Fort Howard (Green Bay), went into business, and in 1874 bought an interest in a mill at Medford which, at the time, was operated by Roberts & Whelen. David had three children by two marriages, and Lizzie, one of his daughters by the second marriage, married T.C. Whelen. McCartney remained in Green Bay while his partner ran the mill in Medford.

Knapp, Stout & Company, the largest sawmill complex in Wisconsin, at Menominie, about 1890. In the foreground, left, is a sawmill called the "water mill" because it operated on waterpower. In left foreground are platforms where lumber rafts were assembled and when finished, were tilted directly into the river. In lower right are the "Tainter Gates," named for Jeremiah Burnham Tainter, who invented a new type of water gate that did not operate vertically, or up and down, but was built something like a rocker, and today used throughout the world. The river spilling through the gates in the foreground is the Red Cedar. The big building in upper right is a shingle mill, straddling Wilson Creek which joins the Red Cedar at this point. A short time after this picture was taken a big slab burner was erected near the stack. At the right of the shingle mill is a house used for kitchen and mess hall, and beyond, a four-story rooming house. The white building in the upper left is the company store where employees used tokens to buy supplies, and beyond the store is a building first used as a recreational center and later as a land office when the company began to dispose of its cut-over lands. At the extreme right, barely visible, is a steam sawmill which was situated on the west shore of the mill pond of Red Cedar River today called Lake Menomin.

Rhinelander, a city of sawmills since its founding in 1882 by the Brown Brothers, perhaps noted with a sigh of sadness the day that this, the last of the big sawmills, shut down on August 27, 1937. The Thunderlake Lumber Company, after whom a lake in Oneida county is named, stood on Lake Creek on the north side of the city where the creek was temporarily dammed up to make a mill pond.

The Rice Lake Lumber Company on Rice Lake, about 1900, lumber yards at left, log boom in center. The two stacks in center foreground probably represent the planing mill, while the main sawmill, with its own water tower, stands in the distance. H.C. Reed took the photograph.

Sawmill at Elton, now a ghost town, which in 1895 had a population of ninety-five people. The village lay about twenty-five kilometers east of Antigo on the Chicago & North Western railway.

The Hines, Alexander & Edgar Lumber Company mill at Iron River (Bayfield County).

Francis Biron, an immigrant from lower Canada, built this mill on the Wisconsin River at "Biron," probably on the site of the first up-and-down mill he built in 1839-40. The mill shown here ran about forty years before it was replaced by the "Grand Rapids Pulp & Paper Company," today Consolidated Papers Inc.

Mill of the Northern Wisconsin Lumber & Manufacturing company on Lake Superior at Ashland. The mill began operations about 1900. From 1911 to 1919 it was run by the Willow River Lumber Company, and in 1919 reverted to its former owners who operated it until it burned, about 1922.

The Davis & Star Lumber Company mill at Rib Lake, about 1910. The strange-looking addition below the stack may be a slab burner, the only one of its kind in Wisconsin. A dam on Rib River formed the mill pond seen in the foreground, and log trains from camps to the east of the village came directly up to the pond on a siding where they were dumped into the water.

Silhouette of the big sawmill at Chippewa Falls for Chippewa Lumber & Boom Company, a wholly-owned subsidiary of the Mississippi River Logging Company controlled by Weyerhaeuser. The mill was alleged to be the largest in the world under one roof and operated on three floors instead of the usual two.

Lumbermen and their families sought surcease from the responsibilities of running sawmills and camps by taking an excursion, like this curious flotilla at Tomahawk, about 1900. Old railway car tops, with windows added, were probably used for sleeping quarters, while the cook and cookee occupied the canvas-covered wanigan at the rear.

One of the largest mills in Marinette County, this one was built by the Goodman Lumber Company in a village named for the Goodman family. The manufactured lumber was shipped out on the Soo Line railroad. The twin bull-slides (right) suggest that there were two saw decks in operation, the only picture where this has been noted. The milk-bottle structure (left) is the slab burner.

The Salsich & Wilson Lumber Company, about 1899, on Star Lake (Vilas County). This company, shortly changed to Williams & Salsich, logged off most of the pine around Star Lake, Laura, Irving, Ballard, Partridge, and northeast to Buckatoban and Black Oak. The sawmill stood on a spit of land running into the lake from the north shore. The Milwaukee Road came in from Woodruff via Sayner to pick up the lumber. On the sidehill at the right are company owned houses for employees.

A "pulp mill," built by Henry Sherry in a village named after him in Wood County. The 1898 *Gazetteer* gives the population of Sherry at one hundred and thirty-eight, and as industry, a sawmill and heading mill. No mention is made of a pulp mill but the word "pulp" here should be translated as "heading mill," a term no longer used, but in the 19th century associated with the manufacture of barrel staves, wooden tubs, and other articles requiring short logs, more commonly referred to as "bolts" or pulp. Mill Creek (left) has been dammed up to make a pond for log storage.

The sawmill at the right was built by Philetus Sawyer who, like Isaac Stephenson, rose from rags to riches in the lumber business and capped his career by winning a seat in the United States Senate. The mill stood on Jewel-Sawyer Creek in Oshkosh, and this building was used from 1853 to 1874. Sawyer arranged to have the first fire engine house built near his mill, since most of the volunteer firemen, no doubt, were his own employees. The fire house seems to be new and has no bell. The volunteers are standing in a line holding what appears to be a long piece of fire hose. Board sidewalks were needed, not in summer time so much, but in the spring and in rainy seasons when roads and paths turned to mud.

Picture postcard (slightly enlarged) postmarked Peshtigo, 1907.

The Merrill *Advocate* on December 23, 1902 had a lead story about the "New Town of Heineman. An Important Industrial Center—A Fine Market Place for the Farmer's Products—A Very Lively Burg." The village, which lay east of Merrill in Lincoln County, was built by Sydney Heineman, president of the National Bank at Merrill, and his son, Harry H. The family held the controlling stock in the Heineman Lumber Company. The *Advocate* reported that the village "consists of a well appointed general store, hotel, drug store, saloon, blacksmith shop and about twenty-five dwelling houses, the homes of contented and happy workman. . . ."

On July 21, 1910, the *Advocate* ran a big headline: "The Worst Holocaust in History of Lincoln County." No lives were lost in the fire but the entire village of Heineman, homes, mills, and yards were wiped out, and never rebuilt.

After sawing and logging around Hazelhurst and Arbor Vitae for two decades, Yawkee-Bissell Lumber Company of Wausau moved its operations to White Lake in Langlade County in 1916, and this photo shows mill at peak of operations in 1920s. It burned February 4, 1950.

One of the few sawmills built on the Wisconsin side of the St. Croix River was this one at Hudson for the Hudson Lumber Company. The first mill was built in 1882 and later enlarged. The mill continued in operation until November 10, 1917 when it was shut down. Approximately four hundred million feet of lumber were manufactured at the mill, and it was also the last mill to operate on the St. Croix River.

Below: Postcard picture (enlarged) of Ogema Lumber Company sawmill at Ogema, dated 1908.

East of Couderay (Sawyer County) a short distance once stood a big mill works on a flowage created by a dam on Eddy Creek. An historical plaque on Highway 70 calls attention to the mill. This one may have been built in 1902 by Levi Felton and P.M. Parker.

The raw atmosphere of the lumberjack frontier is well illustrated in this picture taken at Emerson (near Cable), in 1908. At the left are employee houses, shacks and log cabins. In the foreground stands skidways of logs waiting to be sawed. From the sharp angle of the "bull-slide" on the mill, it seems that the logs were skidded up the slide by a windlass with a chain or cable.

The city of Bayfield lies mostly on a hillside overlooking Lake Superior and Madeline Island to the east. The hills to the west were once spotted with groves of pine, and sawmill operations began at Bayfield in the mid-1850s. In 1870 Robinson D. Pike, referred to in early texts as "captain," built a mill on the waterfront which came to be called the "Little Daisy." In 1905 the Pike mill was absorbed by the Henry W. Wachsmuth Lumber Company. Wachsmuth started as a mill worker in Bayfield, later had his own retail yard, and finally took over the "Little Daisy," which he immediately enlarged. There is not much room for a yard along the Bayfield waterfront and, as this picture shows, slab wood has been used to build up a little lake frontage which made it easier for lake boats to come alongside and pick up cargo. The harbor breakwater, built by Wachsmuth, was donated to the city for a marina after the mill shut down in 1924. Common labor at the mill in 1888 was paid $1.65 to $1.75 a day; the head sawyer $3.50, engineers $3, and the saw filer, the best paid of all, $4 to $5 per day.

KEY TO SAWMILLS IN BAYFIELD COUNTY

1.	Red Cliff Lumber Co.	1898-1905
2.	Bayfield Land Co.	1856- ?
	R.D. Pike Lumber Co.	1869-1912
	Wachsmuth Lumber Co.	1912-1924
3.	A.A. Bigelow Lumber	1886-1902
	Edward Hines Lumber Co.	1902-1905
	Rood & Maxwell Lumber	1885-1887
	South Shore Lumber	1897- ?
	Washburn Box Factory	1897- ?
	Fowler & Jacobs Lumber	1898-1903
	C.C. Thompson Lumber	1889-1907
4.	White River Lumber	1889-1905
5.	Chicago Lumber	1903-1911
6.	Cornberger Lumber	1895-1910
7.	G. Levine Lumber	1883-1900
8.	Niel & Pratt Lumber	1883-1900
	Wells, Ruggles & Jackson	1886-1893
	Stanstrom & Skoglund	1892-1912
9.	J. Campbell Mill	1904-1908
10.	H. Nelson Mill	1915-1922
11.	Pough Mill	1904-1910
12.	Junek Mill	1909-1930
13.	Reed & Nystrom	1909-1930
14.	Good Lumber	1905-1926
15.	Rust-Owen Lumber	1882-1930
16.	Rogan Lumber	1906- ?
17.	Rogan Lumber	? -1910
18.	Radloff Bros. Lumber	1909-1920
19.	Radloff Bros. Lumber	1920- ?
20.	Bairs Lumber	1930-1940
21.	Fink Lumber	1938-1958
22.	Hessy Mill	1906- ?
23.	Phips & Martel	1893-1910
24.	Crystal Lake Lumber	1898-1910
25.	Lea Ingram Lumber	1892-1899
	Alexander Edgar Lumber	1900-1904
	Edward Hines Lumber	1905-1913
26.	Spider Lake Lumber	1899-1902
27.	George Stuntz Mill	1854-1856
28.	Pat Hynes Mill	1892-1897
29.	Ebony Manufacturing	1892-1897
	T.N. Okerstrom	? -1925
30.	Cranberry Lumber	1890-1899
31.	Thompson & Flieth Mill	1908- ?

Ellingson Lumber Company sawmill at Hawkins in northern Chippewa County. Picture is enlarged from steropitican "double take" photo taken in 1906 by C.J. Monroe who had studio in Hawkins. Village got daily mail on Omaha railway.

Aerial view of C.M. Christiansen lumber mill complex at Phelps, about 1950. Main mill stands on North Twin Lake in foreground, with lumber yards at left and chemical factory in upper center. Village of Phelps would be at right of picture.

Interior of C.M. Christiansen sawmill at Phelps, about 1940. Man at left rides carriage and handles set works, pushing lever forward each time log is returned from saw. Man in center is head sawyer who holds hand on lever which controls carriage.

Lumber yard, probably for T.B. Scott Lumber Company at Merrill. Tracks were laid into the yard and a truck was used to bring the sawed lumber to the piles for drying. Lumber pilers, as this picture reveals, took pride in the precision with which the piles were made, tier on tier, inclined slighlty forward to provide maximum drainage. Not a single piece overlaps or seems out of place.

The lumber "jacker" (lower) uses a two by four, inserted into the pile of two by sixes, to use as a fulcrum. He places the plank on the fulcrum and leans down on one end to jack it up to the piler. The latter usually wore a leather apron because he balanced the piece on his pelvis. Without the apron he would wear out a pair of trousers in a day.

The mountains of slab wood that were accumulated at most mills were always a problem. In some yards the slab was laid down as fill, or burned in a steel burner, like the one seen at the right, or hauled away for kitchen stove wood. Here a sign advertises "Order Your Wood of John Lee, the man who hauls wood (probably "home") on time."

Before semi-trailers and trucks, lumber was hauled to town or to railway siding by wagon with front wheels lower than the back wheels. In this picture, the lumber seen here will be loaded into box cars at a siding of the Chicago & North Western line in Rosholt, about 1912. The lumber was hauled eight miles from the sawmill of Alfred Wrolstad in Northland (Waupaca County). The teamsters rigged a spring-board seat for themselves, but the chain holds the load intact. The first teamster on the right is Sam Opsal; second, left, Andrew Meyer; third, Martin Wrolstad; and fourth, Henry Meyer, with Alfred Wrolstad seated (in dark suit) on wagon below Meyer. Others not identified.

Crew of the Ashland Shingle Mill. The logs brought to this mill were mostly cedar which had already been bucked into "bolts" or short pieces about three feet long. These were sawed by machine and made into shingles and tied in bundles (left).

There were no wasted corners in this building at Tomahawk called "Tomahawk House," which catered to mill hands and traveling salesmen. At the left is wood for the coming winter. The enclosed door at the left led to the front desk and usually a long bar, with table for poker and whist at the rear, and possibly, a pool table. The photographer failed to see the rope or hose lying on the board walk. Everyone in the picture appears to be dressed for Sunday.

Lumbermen

Frederick Weyerhaeuser

George N. Paine

Daniel Shaw

Philetus Sawyer

John Week

Edward Hines

Right, a genuine Indian bark canoe, near Menomonie. Seated in the bow is Johnny Coady and in the stern, Alex Moose, two Indians in the employment of Andrew Tainter who sits behind James Bracklin (in white shirt). Tainter, one of the first sawmill operators in Menominie, in the 1850s, took a Chippewa woman to wife and had five children by this marriage before he was divorced and remarried.

One of the early log birling contests was held on the Mississippi River off Alma where Gerhard Gessell shot this picture, about 1890. In the early 1900s, birling contests became an annual event in several cities of the north, most of them held on the 4th of July. At a birling contest held at Ashland on Labor Day, 1901, Tom Oliver, Michigan's best, was pitted against Jim Stewart, Eau Claire's finest. They started to birl after lunch and the contest went all afternoon before called on account of darkness. The next day, three hundred jacks were in town to lay their bets. Stewart won only by outlasting his opponent who collapsed in late afternoon. The winner took $100. Other well-known birlers would include Tom Macaroon, Jack Fleming, Tom Murray, and the Thompson brothers, Abe and Thomas, all of Eau Claire.

Above, T.B. Scott, and upper left, A.H. Stang, leading lumbermen of Merrill. Scott funded the T.B. Scott Free Library at Merrill.

Isaac Stephenson, pioneer lumberman of Marinette, politician and United States Senator from Wisconsin.

Hiram P. Graham (above), was associated with the Dells Lumber Company in Eau Claire and, in the early 1900s, with the Phoenix Manufacturing Company which made steam haulers. He was also the first mayor of Eau Claire.

In center, surrounded by his four sons, is DeVillo R. Holt, who began in the lumber business in Chicago, and in the 1870s moved to Oconto County where he was associated for a time with Uri Balcom in the Holt & Balcom Lumber Company. His sons later took over management of the firm, namely W. Arthur (upper left), Donald R. (upper right), Charles (lower right), and George (lower left).

Almost every lumberjack east of Glidden, south of Exeland and north of Cornell knew Canadian-born Edward LeMay, who made his first log drive in Chippewa County in the 1880s. Later he worked as a jobber, getting out timber to mills for others. He logged on the South Fork and North Fork of the Flambeau, on the Fisher, Mud, Brook, Thornapple, Jump and West Fork of the Chippewa. Many winters he had crews in camp of more than a hundred men and equally as many horses.

John S. Owen, one of the founders of the Rust-Owen Lumber Company at Drummond and director of several large mills in Eau Claire and in Clark County.

Left, Henry Sherry, one of the leading lumbermen in the Wolf River Valley, operated a sawmill in Oshkosh but maintained a large retail yard and office in Neenah where he also lived. He started in business in 1860 with George Rogers running a threshing machine made by Smedley & Case of Oshkosh.

Right, A. (for Anderson) W. Brown, often called "the father of Rhinelander," formed a partnership with his brother, W. (for Webster) E. Brown, to found the Brown Brothers Lumber Company, one of the leading lumber firms in the city for many years. The Browns purchased 1500 acres of timber in the lake country and along the Pelican River to begin lumbering in 1882. This led to the founding of Rhinelander, named for a New York banker who helped raise the bonds to build a railroad to the newly established lumber town.

Right, F.S. Robbins, who came to Rhinelander in 1886, was both a rival and an associate of the Brown Brothers in the lumber business. Logs for his mills were brought in over narrow gauge railroads that went as far as Sugar Bush in Vilas County, and to Pine Lake in Forest County.

C. (for Charles) M. Christiansen (left) acquired the big sawmill of Hackley, Phelps & Bonnell Lumber company at Phelps in 1929. Despite a Depression just around the corner, he survived and enlarged his mill facilities and bought more timberland. The mill closed in 1959 and the assets of the company were taken over by two sons and now administered by P.C. Christiansen. The family originated in Wausau.

Postcard picture of sawmill at Weston (1908) located south of Menomonie in Dunn County. Village, now a ghost town, had daily mail via Chicago, Milwaukee and St. Paul railway which had station nearby.

Sawmill of John Schroeder Lumber Company. Despite silo-looking slab and sawdust burner at left, long piles of firewood lie in foreground, and man in dump cart hauls load away for local customer. Long "bull slide" at left of mill runs into the bay to convey logs to second deck of mill where saws were located.

Postcard picture of Hatten Lumber company sawmill on Wolf River in New London, about 1910. Company originated in Manawa as Lindsay & Hatten.

Postcard picture of the R. Connor Lumber Company sawmill at Laona (1908). Connor built first mill at Stratford in Marathon County. The above mill has been abandoned in favor of a hardwood mill, once largest in the U.S.

The Rhinelander Logging Museum, "Home of the Hodag," stands in a park located off Highway 8 on the south side of the city.

The Public Museum at Wabeno, built in traditional camp style with dingle in the middle, stands on Highway 32 in the village. A steam hauler, the best preserved of any in the state, is on exhibit on the grounds of the museum.

Tobacco cans on exhibit at the Menominee Logging Camp.

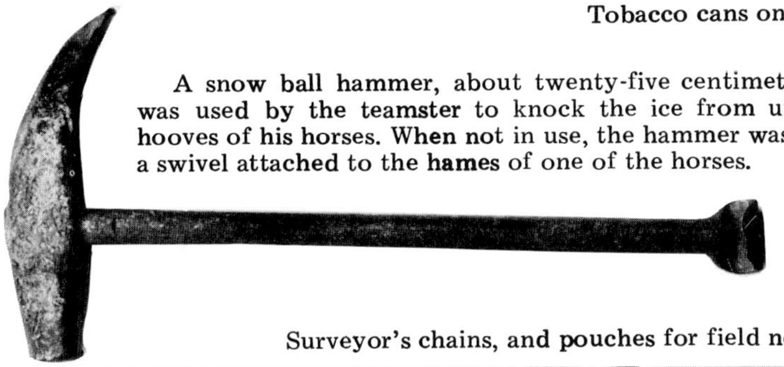

A snow ball hammer, about twenty-five centimeters long, was used by the teamster to knock the ice from under the hooves of his horses. When not in use, the hammer was hung in a swivel attached to the hames of one of the horses.

Surveyor's chains, and pouches for field notes, on exhibit at Menominee Logging Camp.

OSHKOSH TOOLS

Oshkosh Hooks

ALL Oshkosh hooks are made of special crucible round-edge hook steel of high carbon content. The round edge prevents the side of the hook from catching when it is thrown into position. The steel is tough and stiff, not brittle, and the points remain sharp. Hooks are hammer-forged, not drop-forged. They are accurately bent, then ground to a sharp point.

Duck Bill Hook

Round Bill Hook

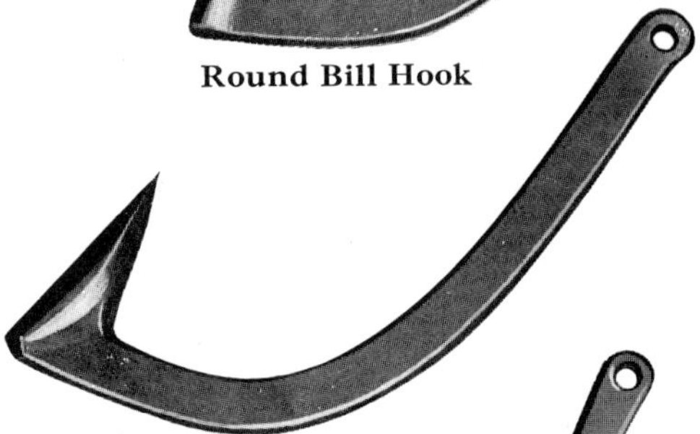

Diamond Head Loaders' Hook

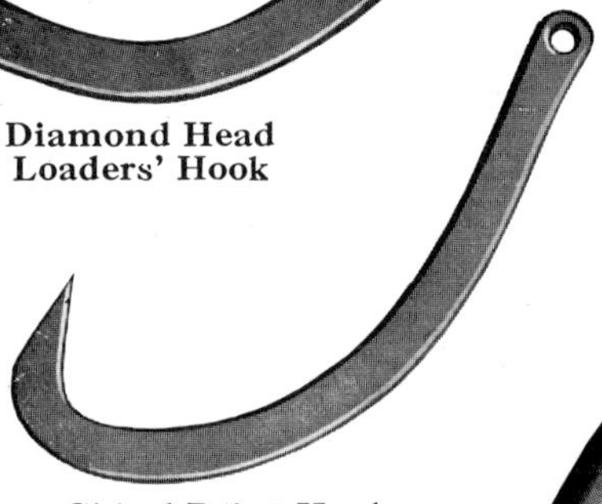

Chisel Point Hook

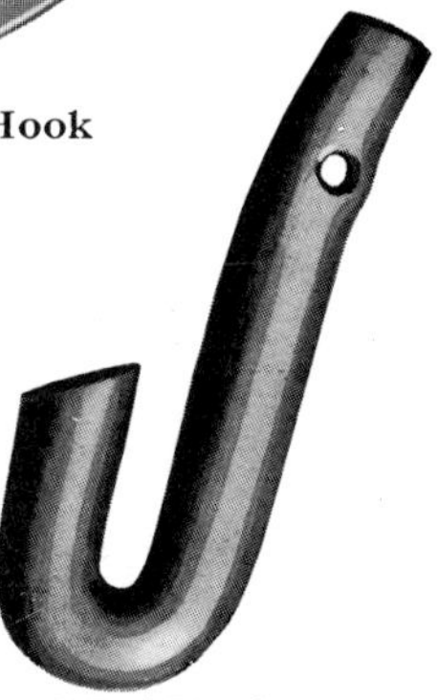

Oval Fid Hook

Flat Fid Hook

Flat Fid Hook with Ring

Oval Fid Hook with Ring

DUCK BILL HOOKS

No.		Weight	Dozen
479	7/8 x 7/16 in.	20 lbs.	$ 9.20
480	1 x 7/16 in.	23 lbs.	10.50
481	1 x 1/2 in.	27 lbs.	12.70
482	1 1/8 x 1/2 in.	35 lbs.	14.80
483	1 1/4 x 1/2 in.	38 lbs.	17.70
484	1 1/2 x 1/2 in.	42 lbs.	21.00

ROUND BILL HOOKS

No.		Weight	Dozen
486	1 x 7/16 in.	23 lbs.	$10.50
487	1 x 1/2 in.	27 lbs.	12.70

DIAMOND HEAD LOADERS' HOOKS

No.		Weight	Dozen
492	1 x 1/2 in.	27 lbs.	$14.70
493	1 1/8 x 1/2 in.	35 lbs.	16.90

CHISEL POINT HOOKS

No.		Weight	Dozen
490	1 x 7/16 in.	22 lbs.	$ 9.50

FID OR CORNER BIND HOOKS

Made of either oval or flat steel bent to the proper shape. They are furnished with or without ring. The hooks are stiff and will not spread. They are neatly forged, are clean and look well.

WITHOUT RING

No.		Weight	Dozen
451	1 1/4 x 5/8 oval steel	22 lbs.	$ 4.90
452	1 1/4 x 3/8 flat steel	22 lbs.	4.30
453	1 1/2 x 1/2 flat steel	28 lbs.	7.60

WITH RING

No.		Weight	Dozen
454	1 1/4 x 5/8 oval steel	24 lbs.	$ 7.00
455	1 1/4 x 3/8 flat steel	24 lbs.	6.40
456	1 1/2 x 1/2 flat steel	28 lbs.	10.60

From Oshkosh Tool Company catalog of 1939.

PEAVEY

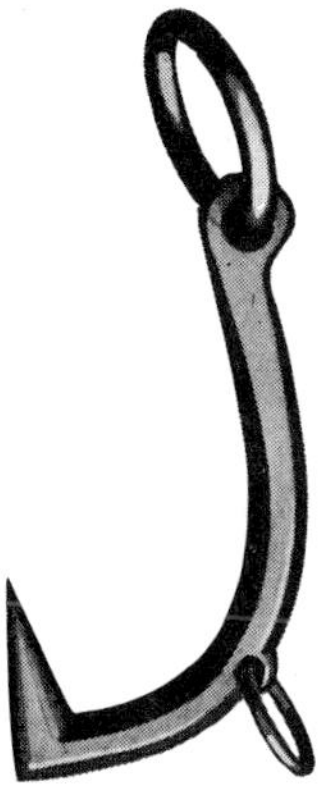

Wire Rope
Loading Hook

Loading
Hook

Lug Hooks

Swamp
Hook

Grapple Hook

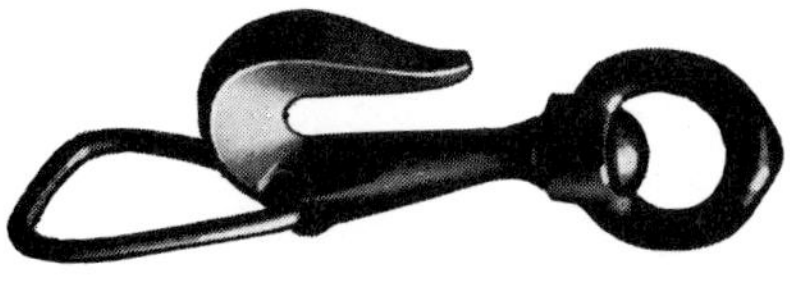

Evener Swivel Hook

Saw Wedges

Bark
Spud

Hookaroon

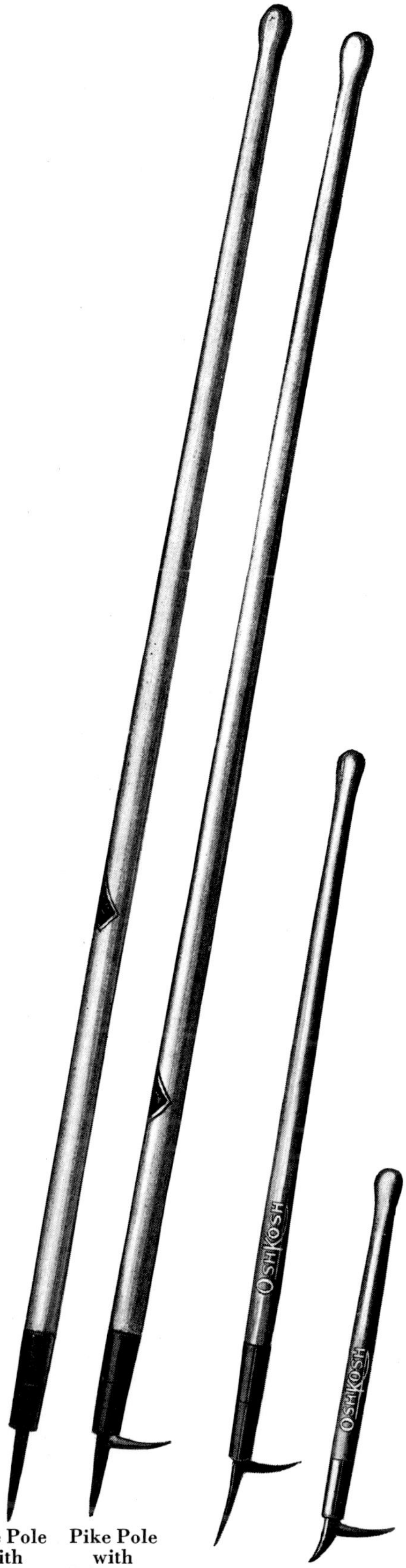

Pike Pole
with
Pike
Only

Pike Pole
with
Pike and
Hook

Mill
Pickaroon

Lumber
Sorters'
Pickaroons

HENRY SHERRY,

MANUFACTURER AND DEALER

Pine and Hardwood Lumber, Timber, Lath,

SHINGLES, PICKETS, CEDAR POSTS, ETC.

MILL AND YARD,
NEENAH, WISCONSIN.

MILL AND YARD,
FOND DU LAC, WISCONSIN.

Neenah, Wis. July 14—1880

Mess Holt & Balcam
Gents
I have sold 300 m
of the Sec 33 logs to Radford & Bro
at 8.00 per m on one year
time at 7 % with Bray & Choate
endorser, this is the best offer
I have had considering the paper
and I think it is all they are worth.
they take about 70 M and I should
think 1/5 norway. I have sawed one
raft to try them and I think it
is all they are worth, shall sell
the ballance at same price to "R"
if they will take them, and you are
satisfied with sale. It is bad
time to have logs at boom. Had
two severe blows and have broke
up bad, shall want to get them
off my hands at once, let me
hear from you at once
Yours &c
Henry Sherry

The house that Henry Sherry built on South Wisconsin Avenue in Neenah was an ornate example of Victorian Eclectic. Despite its wooden ornamentation, it is a brick structure, whereas most Victorian Eclectic were frame. The four-story tower has a Mansard roof pierced by Gothic, gabled dormers, and three Roman-arched windows on each side. In back is probably an old horse barn converted to a garage. The picture was taken in 1979.

Francis Biron built his house in Gothic Revival. It stood near his sawmill on the Wisconsin River and today is used as an office of Consolidated Papers Inc., Biron Division. The architectural style of the house suggests that it was built some time between 1859 and 1872.

The Cook-Rutledge Mansion at Chippewa Falls, now a public museum.

The greening of Wisconsin

The resurrection of Wisconsin forest lands after a century of intensive logging has been a long, uphill strugggle between justifiable human need and unjustifiable human greed. The movement to revive the "greening of Wisconsin" was sparked by Edward Merriam Griffith, Wisconsin's first state forester and one of the first in the United States with European forestry training. But when he urged, in his annual report for 1909-10, that counties be empowered to acquire forests in cooperation with the State Board of Forestry, he was met with open hostility. The issue seemed to approach a dead end when, in 1915, the State Supreme Court declared that most of the forestry laws passed by the legislature were unconstitutional. This was the final blow to Griffith who, deeply disappointed, resigned and left Wisconsin.

Oneida County took the lead in opposing the forestry program proposed by Griffith and in 1913, the county board commended members of the legislature who had opposed the state forestry program "for the good work they have done in Madison, toward putting an end to Mr. Griffith's pet scheme to gobble up our best agricultural lands."

In short, it was still the will of the people, even though most of the pine timber was already gone, to do nothing towards reforestation or towards calling a halt to indiscriminate timber cutting, while encouraging cultivation of land no matter how sandy, stony, swampy, or unsuited to farming it was.

Twenty years later Oneida County took the lead in passing zoning regulations and promoting county forestry in Wisconsin. Today, county forests in the state total more than 2,200,000 acres, or more than the sum of state and national forest lands.

Although no one ever counted the pines that were harvested in Wisconsin before 1900, there are probably more growing in the state today than there were a century ago, all planted by hand with a spud or by machine planter.

Meanwhile, Griffith was instrumental in establishing a tree nursery at Trout Lake (Vilas County), and in 1912 red pine seedlings from the Trout Lake nursery were planted at Star Lake, the first tree plantation in the state.

However, the tree planting program made little headway after Griffith left the service, and it was not until World War II, when private industry, mainly the paper mills, pressed by the prospect of future shortages of pulp, helped to found Trees for Tomorrow which was incorporated at a meeting of state foresters and representatives of industry at Wausau, February 29, 1944. A headquarters compound was built at Eagle River for the organization, and here educational programs and promotion work was carried out to expand on the idea of planting and cultivating trees. The program was taken up by high school students throughout the north and today there are numerous school plantations growing timber, halting erosion, and providing areas for environmental studies and recreation.

By the late 1960s, Trees for Tomorrow was running out of fields and old clearings for tree planting, and the program was terminated in the early 1970s. The emphasis is now on environnmental studies and education. During the span of the program, Trees for Tomorrow, according to M.N. Taylor, its long-time executive director, distributed and machine-planted 23,000,000 trees, prepared management plans for 370,000 acres of private woodland, and estimated, marked and supervised the sale of 440,000 cords of wood.

However, the "greening of Wisconsin" would not have taken off so easily had it not been for the groundwork laid by the Civilian Conservation Corps in the 1930s. This organization was authorized by President Roosevelt on May 5, 1933, not long after he took office. Across the nation there were thousands of young men who were unemployed in the Great Depression, and the Civilian Conservation Corps was created to give them temporary employment and physical training. Army-type camps were built in the woodlands and mountains where the men worked in tree planting, water conservancy, erosion control, dam building, riprapping of rivers and sloughs, cutting fire lanes and, in general, assisting in the "greening of America" as well as of Wisconsin.

A milestone was reached in the Trees for Tomorrow program when the ten-millionth tree was planted for a private landowner on April 29, 1962. To the right, M.N. ("Mully") Taylor, executive director of Trees for Tomorrow. Seated on the planter is Floyd Rhode, and facing him, Robert Engelhard, forester. The plow on the planter makes a furrow and when the seedlings are dropped into the furrow, the two rubber wheels pack them firmly into the ground (lower left). The seedlings, about the length of a man's hand, are carried in two iron boxes at the front of the machine where the person on the planter, can grab a handful quickly and insert them in the furrow, one at a time, without halting the momentum of the driver on **the tractor. The seedlings here appear to be red pine.**

The Corps was headed up by United States Army personnel. Olive drab uniforms were issued to the recruits and the pay was a dollar a day and board. People called them the "C.C.'s." They did a remarkable job cleaning up the slashings in the cut-over lands, planting trees and helping to control forest fires.

The "C.C." camps began to fold after war broke out with Japan in December, 1941, and by the spring of 1942, only a handful of camps were still active. On July 30, 1942, the Corps was deactivated. By that time the majority of the boys had either volunteered for military service, or were being drafted. Their training with the "C.C.'s" gave them an advantage over other recruits because they understood the meaning of discipline and duty. More than two million young men were engaged at one time or another in the program. Although they did not realize it at the time, the "C.C.'s" were contributing to the "land ethnic" of Aldo Leopold who looked on the land as a community to which we all belong, not a commodity belonging to us.

...cers

Camp View

Company 2622

John S. Kapuscinski, ...
Raymond B. Steiner, 2nd Lt. Cav-Res. Exe...

...reation Hall

Mess Hall

Camp Delta, Drummond, Wis.

Foresters

-V (G) USNR, Commanding
L. A. Wangard, Camp Superintendent
Photo Spencer

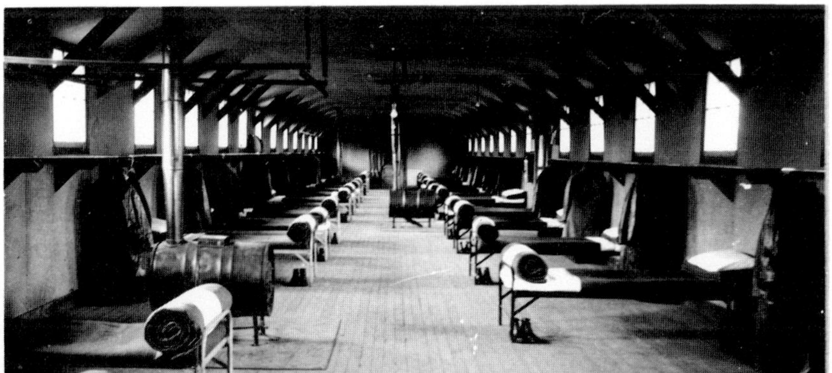

Barracks

Retreat

Trees for Tomorrow camp at Eagle River.

Frederick G. Wilson was along when the first trees were planted at Star Lake in 1912 by foresters of the Wisconsin Conservation Department. He later served as a state forester and before his retirement in 1952, as superintendent, forest management division.

This settler was close enough to big timber to build his own log house. Roof is covered with shakes, probably shaved from cedar bolts. Eave troughs hang precariously at corners of house but aimed at rain barrel. Balsam tree seems propped into turf to cover up lean-to entry built to protect back door from wind. Behind house at left is also lean-to, and at right, log stable with rail fencing. Small stovepipe on gable suggests only cooking stove or range in main part of house. This picture surfaced in Iola, Wisconsin, 1984. On the back appear these words (in parenthesis) "Mary Frogner" and below, "Nannes house in Dorchester." The Frogners were a large family in Iola in the 1880s and 1890s, and it seems that one moved to Dorchester, in Clark County (north of Abbotsford). Style of dress on mother and children suggests picture taken in 1890s.

Log end marks

(Reprinted by special permission of Howard Peddle from *Wisconsin Log Marks*, 1980.)

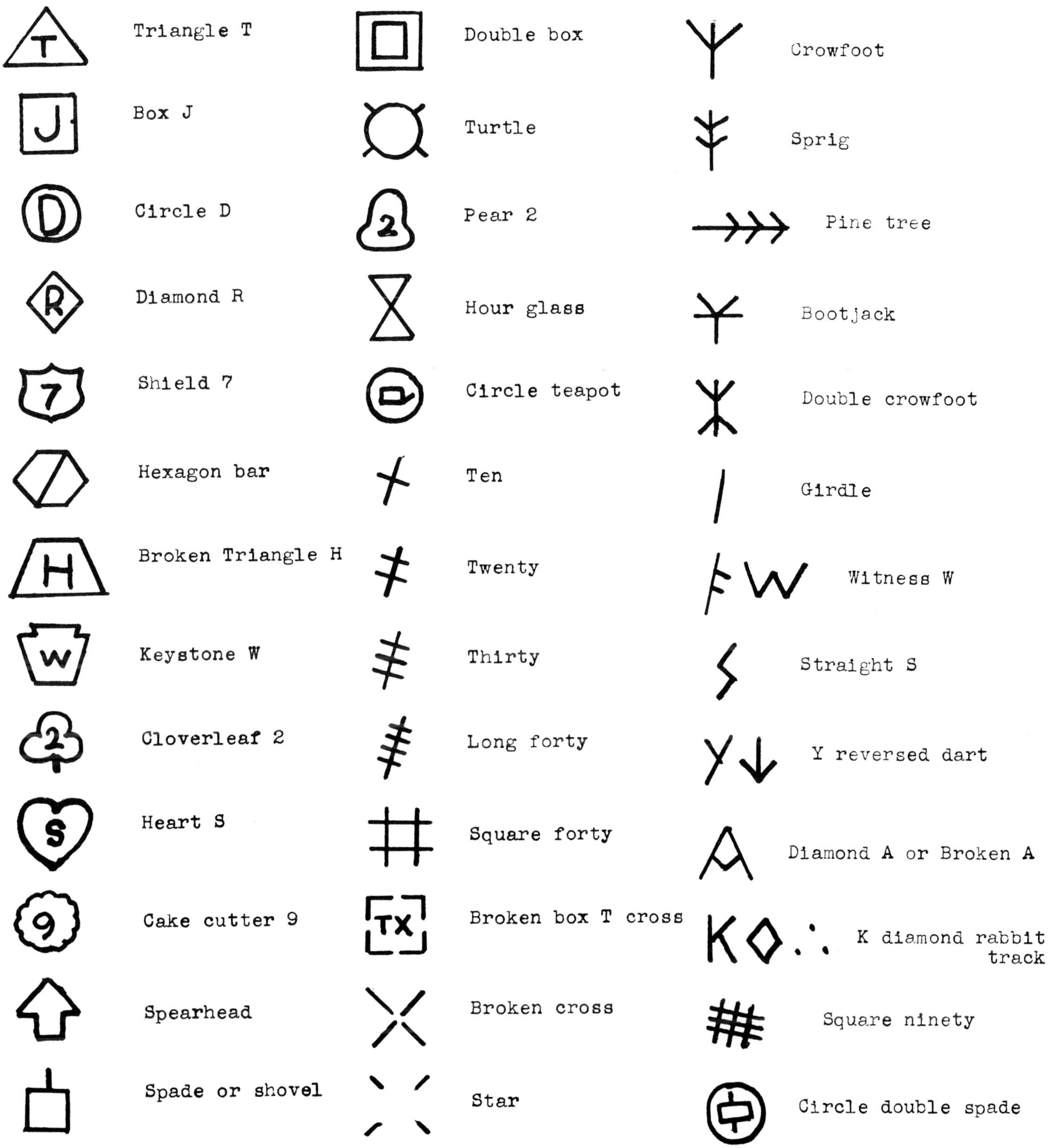

Log End Marks from Ashland Lumber District, 1887-1908

End Mark	Registered By	Date of Registry
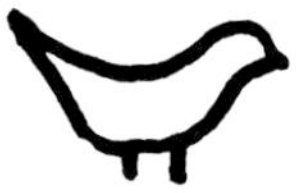	E. J. Waters	5 Dec., 1887
B&B	F. Boutin Sr. & F. Boutin Jr.	18 Apr., 1888
	C. C. Thompson & Walkup Co.	13 Nov., 1888
CITY	Wright & Ketcham	7 Jan., 1889
94	Cranberry Lumber Company	15 Oct., 1889
	Sand River Lumber Co.	10 Apr., 1891
	Bernard Doherty	2 Nov., 1891
MA	South Shore Lumber Co.	27 Dec., 1892
	Patrick Hynes	18 Nov., 1893
JBM	J. B. Moore & Co.	21 Dec., 1893
	North Wisconsin Lbr. Co.	6 Oct., 1894

End Mark	Registered By	Date of Registry
	Theodore Okerstrom	26 Nov., 1894
MH	Mark Hessey	20 Dec., 1894
ASH	A. J. Mussell	25 Jan., 1895
CBS	C. B. Simpson & Co.	1 Nov., 1895
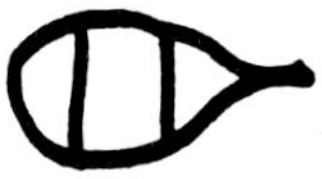	John A. Jacobs	19 Dec., 1895
	E. G. Filer (Manistee, Mich.)	19 Oct., 1896
	Weyerhaeuser & Rutledge	16 Feb., 1897
	W. H. Gilbert	28 Oct., 1897
	Red Cliff Lumber Company	30 Nov., 1897
JOB	John O'Brien Lbr. Co.	22 April, 1898
R	Rittenhouse & Embree	4 Oct., 1898

A Brooks 8

Brooks & Ross camp, about ten miles southwest of Antigo, probably winter of 1909. A hundred men worked in this camp and the teamsters shown here are holding more than fifty horses. Two men in white aprons, center left, atop an icing tanker, are cooks. The cook shanty and bunk camp are in rear at left, the horse barns and blacksmith shop at right. At far left is a second icing tanker. Logs cut here were shipped out via the Gleason line to Otis on the Chicago, Milwaukee & St. Paul railroad, thence south to the company mill at Schofield. Joe Wanta from Rosholt was woods boss that winter.

Ross camp

Monsters of the past

The 19th century was marked by sporadic discoveries of ancient giants and strange animals which, when investigated, turned out to be hoaxes. In Wisconsin, Gene Shepard, while out hunting one time, came across a new creature (right) and vowed it was no hoax. He called it a "Hodag" and after its capture, he had it mounted for the Rhinelander Logging Museum.

There was also the "Highbehind," related, it is said, to the Hodag on its mother's side, and through his father, a second cousin to the Sidehill Gouger. A George W. Doyle, in 1871, allegedly got a glimpse of it in the Lake Superior country, possibly as far west as Rainy River. The illustration shown here is merely an artist's conception of the Highbehind because, like the Snow Leopard of Nepal, it successfully evades hunters and camera people. It is said that the Highbehind, hanging from a section line (note section line atop sketch), over a trail in the woods, catches its prey and squeezes it to death. But the Highbehind could not stand clearings, and as the lumbermen cut deeper and deeper into its habitat, its numbers dwindled, and by 1900, according to an old logger who had talked to Paul Bunyan's walking boss, one of the last of its species was seen back tracking up Ding Dong Creek off the east fork of the Chippewa in Section 37. That was a long time ago though.

PICTURE CREDITS

4—Consolidated Papers Inc., Wisconsin Rapids
6 & 7—Author's Collection
9—Author's Collection
11—Author's Collection
12—Minnesota Historical Society (hereafter as MHS).
13—*Harpers New Monthly Magazine* (hereafter as *Harpers*), (March 1868)
14—Author's Collection
15—Author's Collection
16—A, *Harpers*, (March 1860)
B, MHS
17—A, MHS
B, Mabel Tainter Memorial Library, Menomonie
18—Author's Collection
19—Tigerton Lumber Company, Tigerton
20—MHS
22—Author's Collection
24—Author's Collection
25—Runk Collection, MHS
26—A, Courtesy Dr. R.C. Brown, Eau Claire
B, Courtesy Howard Peddle, Iron River
27—Author's Collection
28 & 29—Author's Collection
31—A, Author's Collection
B, Courtesy C. Luke Leitermann, Pickerel
32—Courtesy Howard Peddle
33—A, Courtesy Howard Peddle
B, Courtesy Laird E. Ross, Buffalo City
34—A, Author's Collection
B, MHS
35—A, Washburn County Historical Society, Washburn
B, Courtesy Howard Peddle
36—A, Author's Collection
B, Courtesy Michael F. Slasinski, Saginaw, MI
38—Courtesy Howard Peddle
39—A, MHS
B, Edward Dobbe, Rosholt
40—Author's Collection
41—Author's Collection
43—MHS
44—A, Courtesy Mrs. Olga Jacobson, Iola
B, Courtesy Mrs. John Henkelman, Merrill
45—A, Eau Claire Public Library
B, Douglas County Historical Museum, Superior (hereafter as DCHM)
46—Public Museum, Ashland
47—A, Courtesy Mrs. George Severson, Waupaca
B, Author's Collection
C, MHS
48 & 49—Author's Collection
50—Courtesy Arthur Hase, Wausau
51—A, Courtesy Harry Curran, Rib Lake
B, Mabel Tainter Memorial Library
52—Author's Collection
53—A, Courtesy Fred Vollmer, Tigerton
B, Author's Collection
55—A, T.B. Scott Free Library, Merrill
B, Author's Collection
56—A, Courtesy Michael F. Slasinski
B, Courtesy Jack Bazille
57—A, Mabel Tainter Memorial Library
B, Author's Collection
58—Courtesy Fred Vollmer
B, Tigerton Lumber Company
59—Courtesy Harry Curran
61—Courtesy Ed LeMay Collection, Cornell
62—A, Author's Collection
B, Courtesy Ed LeMay Collection
63—Author's Collection
64—A, Author's Collection
B, SHSW
65—A, Author's Collection
B, Courtesy Ted Schalfke, Wabeno
66—A, Courtesy Fred Vollmer
B, Marinette County Historical Museum, Marinette
67—Marinette County Historical Museum
68—A, Courtesy C. Luke Leitermann
B, Courtesy Harry Curran
C, Courtesy C. Luke Leitermann
69—A, Iron County Historical Society, Hurley
B, Courtesy William Guolee, Gleason
70—A, Courtesy Harry Heinemann, Jr., Wausau
B, Courtesy Rudy Mueller, Rhinelander
71—A, DCHM
B, Courtesy Lester Peterson, Rosholt
72—A, Courtesy Mrs. Rud. D. Hennig, Bowler
B, Author's Collection
73—A, Kinsbury Photo Shop, Antigo (hereafter as Kingsbury)
B, Courtesy George's, Boulder Junction
C, Courtesy Joy McGee Heitmann, Savannah, GA
74—A, Author's Collection
B, Courtesy Robert Nitzsche, Wausau
75—A, Author's Collection
B, Courtesy Robert Nitzsche
76—A, Courtesy Gordon Sorenson, Drummond
B, Courtesy Mr. & Mrs. Gordon Mortenson, Aniwa
77—Courtesy Jack Bazille
78 & 79—Author's Collection
80—Courtesy UW-LaCrosse
81—A, Author's Collection
B, Courtesy Vesper Holtz, Tomahawk
82 & 83—MHS
84—DCHM
86—Author's Collection
87—DCHM
88 & 89—DCHM
90—A, MHS

B, Courtesy William Guolee
91—Author's Collection
92—Langlade County Historical Society, Antigo
93—Courtesy T.J. Thompson, Bruce
94—A, Courtesy Vernon Baltus, Marshfield
B, Eau Claire Public Library
95—Kingsbury
96—Courtesy Arthur Hase
98—MHS
99—Author's Collection
100—Kingsbury
101—State Historical Society of Wisconsin (hereafter as SHSW)
103—MHS
05—Courtesy Fred Vollmer
106—T.B. Scott Free Library
107—Mabel Tainter Memorial Library
109—A, Courtesy P.C. Christiansen, Phelps.
B, Courtesy, Lucille Alft, White Lake.
110 & 111—Courtesy Gerald Johnson and Krause Publications
112—A, Mabel Tainter Memorial Library
B, Courtesy Harry Curran
113—Mabel Tainter Memorial Library
114—A, Courtesy Harry Curran
B, Courtesy David Jankoski, Stanley
115—A, Courtesy Fred Vollmer
B, Author's Collection
116—Kingsbury
117—A, Author's Collection
B, Eau Claire Public Library
C, Author's Collection
118—Author's Collection
121—SHSW
122—A, *Harpers* (March 1860)
B, ARC—UW-La Crosse
123—MHS
124—A, MHS
B, *Harpers* (March 1868)
125—Author's Collection
126—SHSW
127—A, Historyland, Hayward
B, Courtesy George Blanchard, Cornell
128—Courtesy Randall Rohe, UW-Green Bay
130—Kingsbury
133—Kingsbury
134—Kingsbury
135—Kingsbury
136—Kingsbury
137—Kingsbury
138—A, Kingsbury
B, Courtesy Jacque Vallier, Milwaukee
139—Author's Collection
140 & 141—Author's Collection
142—A, Courtesy Edward Weinmann, Shawano
B, Author's Collection
143—Courtesy Randall Rohe
145—Courtesy Randall Rohe
146—Oshkosh Public Library
147—Oshkosh Public Library
148—Courtesy Ray Specht and R.T.L. Eggate, UW-Stevens Point
151—Courtesy Roy Martin, Winter
152 & 153—S.C. Sargent Collection, MHS.
155—A & B, Chippewa Falls Public Library
154—Courtesy Roy Martin, Winter
156 & 157—SHSW
158—SHSW
159—Author's Collection
160—A, ARC—UW-La Crosse
B, Author's Collection
161—Courtesy Randall Rohe
162—Historyland
163—A, Otto J. Erspamer Collection, Iron County Historical Society, Hurley
B, Author's Collection
164—Mabel Tainter Memorial Library
165—A, Mabel Tainter Memorial Library
B, Courtesy Harry Curran
166—Courtesy Harry Curran
167—Courtesy Harry Curran
169—Author's Collection
170—Gerhard Gesell Collection, SHSW
172—Gerhard Gesell Collection, SHSW
173—SHSW
174—A, ARC—UW-La Crosse
B, SHSW
176—A, Mabel Tainter Memorial Library
B, ARC—UW-La Crosse
177—Gerhard Gesell Collection, SHSW
178—Consolidated Papers Inc.
179—Consolidated Papers Inc.
180—ARC—UW-La Crosse
182 & 183—ARC—UW-La Crosse
184—MHS
185—Author's Collection
186—Author's Collection
187—A, Author's Collection
B, Courtesy Dr. R.C. Brown
188—Author's Collection
189—Courtesy Randall Rohe
191—*Harpers* (December 1855)
192—*Harpers* (March 1860)
193—Marathon County Historical Society, Wausau
194—A, University of Pittsburgh Library
B, Courtesy Walter Leuba and Robert McCullough, Pittsburgh, PA
196—Eau Claire Public Library
197—Dunn County Historical Society, Downsville
198—A, Marathon County Historical Society

B, Consolidated Papers Inc.
201—H.H. Bennett Studio, Wisconsin Dells
203—H.H. Bennett Studio
204—H.H. Bennett Studio
206—H.H. Bennett Studio
208—A, SHSW
B, Eau Claire Public Library
210—H.H. Bennett Studio
211—A, Mabel Tainter Memorial Library
B, Consolidated Papers Inc.
212—A, SHSW
B, Courtesy Robert Fromholz, Ashland
214—The Smithsonian Institution
215—Author's Collection
216—*Pictorial History of the United States* (Hartford, CT, 1847)
217—A, & B, *Dell Artificiose Machine*, Rare Book Division, Library of Congress
C, *Virginia's Discovery of Silke Worms. . .Together with the Making of the Saw-Mill*, Rare Book Division, Library of Congress
218—Courtesy John M. Russell, Menomonie
219—Courtesy Randall Rohe
220—A, Author's Collection
B, The Smithsonian Institution
221—Author's Collection
222—*Fourteenth Annual Report of the Indiana State Board of Agriculture* (1872)
223—Author's Collection
225—*Case Wood Engravings* (1942)
226—Oshkosh Public Library
227—Chippewa Valley Museum, Inc., Eau Claire
228—*American Mechanical Dictionary* (1876)
230—Courtesy Lester Peterson, and R.J. Kellett, UW-Stevens Point
231—*Small Sawmill Operator's Manual*, U.S. Department of Agriculture (1952)
233—Courtesy Hazel and Edith Frederickson, Star Lake
234—Courtesy Randall Rohe
235—A, *Small Sawmill Operators Manual*
B, Author's Collection
236—Courtesy Randall Rohe
237—A, Marinette County Historical Museum
B, *A History of Northern Wisconsin* (1881)
238 & 239—Courtesy Ann Borreson, Sparta.
240—Courtesy E.H. Allington, Stratford
241—Courtesy C. Luke Leitermann
242—Author's Collection
243—T.B. Scott Free Library
244—ARC—UW-La Crosse
245—ARC—UW-La Crosse
246—A, Kingsbury
B, Eau Claire Public Library
247—A, Courtesy Melvin Knoke, Appleton
B, *History of the Glidden Four-Town Area* (ca 1977)
248—A, Courtesy Mrs. Rud. D. Hennig
B, Eau Claire Public Library
249—Courtesy Arthur Hase
250—Neville Museum
251—Courtesy John M. Russell
252—A, Courtesy Rudy Mueller
B, SHSW
253—A, Kingsbury
B, Courtesy Howard Peddle
254—A, Consolidated Papers Inc.
B, SHSW
255—A, Courtesy Harry Curran
B, Chippewa Falls Public Library
256—A, Courtesy Mrs. John Wipperfurth
B, Neville Museum
257—A, Author's Collection
B, Consolidated Papers Inc.
258—A, Oshkosh Public Museum
B, Courtesy David Jankoski
259—A, Courtesy Virginia Heinemann, Wausau
B, Courtesy, Kayle Wahleithner, White Lake.
260—A, Runk Collection, MHS.
B, David Jankoski.
261—Eau Claire Public Library
262—A, SHSW
B, Bayfield Historical Society, Bayfield
263—Courtesy Gordon Sorenson
264 & 265—Courtesy P.C. Christiansen.
266—A, T.B. Scott Free Library
B, Author's Collection
267—A, Eau Claire Public Library
B, Author's Collection
268—A, Otto J. Erspamer Collection, Iron County Historical Society
B, Courtesy Joy McGee Heitmann
269—*History of the Lumber and Forest Industry of the Northwest* (1898)
270—A, Mabel Tainter Memorial Library
B, SHSW
271—A & B, T.B. Scott Memorial Library
C, Marinette County Historical Museum
D, Eau Claire Public Library
272—A, Courtesy Ed LeMay Collection
B, Courtesy Arthur Balcom, Oconto
C, Courtesy George Crowell, Neenah
D, Author's Collection
273—A & B, Author's Collection
C, Courtesy, P.C. Christiansen.
274—Courtesy David Jankoski.
275—Courtesy David Jankoski.
276—Author's Collection
277—Author's Collection
278—Langlade County Historical Society
279—Langlade County Historical Society
280—ARC—UW-Green Bay

281—Author's Collection
283—Courtesy M.N. Taylor, Merrill
284 & 285—Courtesy Gordon Sorenson
286—A & B, Courtesy Frederick G. Wilson, Green Valley, AZ
C, Author's Collection.
287-289—Courtesy Howard Peddle
290 & 291—Courtesy John Mortenson, Aniwa
292—A, Rhinelander Logging Museum
B, Courtesy Tony Wise, Hayward

BIBLIOGRAPHY

Alft, William, "History of Logging Era on Wolf River." Typescript, Oshkosh Public Museum, n.d.

Beef Slough Rafting Works, An Era in the History of Alma. Alma Historical Society, n.d.

Benson, Barbara, *Logs and Lumber: The Development of the Lumber Industry in Michigan's Lower Peninsula,* 1837-1870. Ph.D. dissertation. Bloomington: Indiana University, 1976.

Boyd, Robert E., "Up and Down the Chippewa River." Madison: *Wisconsin Magazine of History,* March, 1931.

Burke, Fred C., *Logs on the Menominee River: The History of Menominee River Boom Company.* Neenah, WI: Banta Printing, 1946.

Connor, Mary Roddis, *A Century With Connor Timber.* Wausau, WI 1972.

Corrigan, George A., *Calked Boots and Cant Hooks.* Saxon, WI 1976.

Current, Richard N., *Pine Logs and Politics: A Life of Philetus Sawyer,* 1861-1900. Madison: State Historical Society of Wisconsin, 1950.

Daily Journal, Little Falls Dam, 1882-1890. Chippewa Falls, WI: Chippewa County Historical Society, 1980.

Dunn, James Taylor, *The St. Croix.* New York: Holt-Rhinehart and Winston, 1965.

Ellis, Albert G., "Upper Wisconsin Country." Madison: *Wisconsin Historical Collections,* volume 3, 1857.

Epler, B.H., *80 Years in God's Country.* Three Lakes, WI 1973.

Everest, D.C., "Reappraisal of the Lumber Barons." Madison: *Wisconsin Magazine of History,* Autumn, 1952.

Fitzmaurice, John W., *The Shanty Boy.* Cheboygan, MI 1889.

Fries, Robert F., *Empire in Pine: The Story of Lumbering in Wisconsin,* 1830-1900. Madison: State Historical Society of Wisconsin, 1951.

Gates, Paul W., *The Wisconsin Pine Lands of Cornell University: A Study in Land Policy and Absentee Ownership.* Madison: State Historical Society of Wisconsin, 1965.

Glover, Wilbur H., "Lumber Rafting on the Wisconsin River." Madison: *Wisconsin Magazine of History,* December 1941, March 1942.

Goethem, Larry Van, *Not Long Ago.* Antigo, WI: The Langlade County Historical Society, 1979.

Gould, John, "Where to See a Regular Lombard." *Ford Times,* September, 1977.

Hall, Walter, *Barnacle Parp's Chain Saw Guide.* Emmaus, PA: Rodale Printing, 1971.

Hallock, Charles, "Life Among the Loggers." *Harpers New Monthly Magazine,* March, 1860.

Helgeson, Thor, *Fra Indianernes Lande.* Iola, WI ca. 1915.

Holt, William Arthur, *A Wisconsin Lumberman Looks Backward.* Oconto, WI 1948.

Holbrook, Stewart, *Holy Old Mackinaw.* New York: The McMillan Company, 1938.

Hotchkiss, George W., *History of the Lumber and Forest Industry of the Northwest.* Chicago: 1898.

Huston, Harvey, *The Roddis Line.* Winnetka, IL 1972.

—*The Thunder Lake Narrow Gauge.* Winnetka, IL 1970.

Ingram, Orrin Henry, *Autobiography.* Eau Claire, WI 1912.

Kenneberg, Adolph, *Log Driving and the Rafting of Lumber.* Wisconsin Public Service Commission, Madison, WI 1944.

Knapp, Henry E., "The Trails from Lake Pepin to the Chippewa." Madison: *Wisconsin Magazine of History,* September, 1920.

Kuony, John, "Lusty Days of Lumber." (Chap. XXVIII, *Prairie, Pines, and People,* Metz, James I., ed., Oshkosh, WI 1976.

Lapham, Increase A.,; Knapp, Joseph G.; and Crocker, Hans, *Report on the Disastrous Effects of the Destruction of Forest Trees Now Going on So Rapidly in the State of Wisconsin,* Madison, WI 1867.

Larsen, Agnes M., *History of the White Pine Industry in Minnesota.* Minneapolis: University of Minnesota Press, 1949.

Latham, Bryan, *Timber, A Historical Survey of Its Development and Distribution.* London: George G. Harrap & Co., Ltd., 1957.

Lincoln, Ceylon C., "Personal Experiences of a Wisconsin River Raftsman." Madison: State Historical Society of Wisconsin, *Proceedings.* 1910.

Lundsted, James E., "Log Marks, Forgotten Lore of the Logging Era." Madison: *Wisconsin Magazine of History,* Autumn, 1955.

Marple, Eldon, *The Visitor Who Came to Stay.* Hayward, WI 1971.

Moran, Joe A., "When the Chippewa Forks Were Driving Streams." Madison: *Wisconsin Magazine of History,* June, 1942.

Nelligan, John Emmett, *The Life of a Lumberman,* as told to Charles M. Sheridan, 1929, n.d.

Olsen, T.V., *The Rhinelander Story.* Rhinelander, WI n.d.

Peterson, Dale, *Lumbering on the Chippewa: The Eau Claire Area,* 1845-1885. Ph.D dissertation. Minneapolis: University of Minnesota, 1970.

Raney, William R., "Pine Lumbering in Wisconsin." Madison: *Wisconsin Magazine of History,* September, 1935.

Rector, William Gerald, "Working With Lumber Industry Records." *Wisconsin Magazine of History,* June, 1950

—"*The Development of Log Transportation in the Lake States,*" 1840-1918. Unpublished Ph.D dissertation, University of Minnesota, 1950.

Reynolds, A.R., *The Daniel Shaw Lumber Company: A Case Study of the Wisconsin Lumbering Frontier.* New York: New York University Press, 1957.

Rohe, Randall Eugene, *Lumbering's Impact on the Landscape of the Wolf River Area of Northeastern Wisconsin.* M.A. thesis. Greeley: University of Northern Colorado, 1971.

Schubert, W.W., "The Menominee River." Madison: *Wisconsin Magazine of History,* Winter, 1950.

Sherman, Simon A., "Lumber Rafting on the Wisconsin River." Madison: State Historical Society of Wisconsin, *Proceedings,* 1918.

Springer, J.S., "Joys and Perils of Lumbering." *Harpers New Monthly Magazine*, March, 1951.
Stephenson, Isaac, *Recollections of a Long Life*. Chicago, 1915.
Stevens, James, *Paul Bunyan*. 2nd ed. New York: Alfred A. Knopf, 1966.
Stoveken, Ruth, "The Pine Lumberjacks in Wisconsin." Madison: *Wisconsin Magazine of History*, March, 1947.
Thorpe, T.B., "*Remembrances of the Mississippi.*" *Harpers New Monthly Magazine*, October, 1855.
Turner, J.M., "Rafting on the Mississippi." Madison: *Wisconsin Magazine of History*, December 1939; March, 1940.
Tuttle, J.M., "The Minnesota Pineries." *Harpers New Monthly Magazine*, March, 1868.
Twining, Charles E., "Plunder and Progress: The Lumbering Industry in Perspective." Madison: *Wisconsin Magazine of History*, winter, 1963-1964.
—*Downriver: Orrin H. Ingram and the Empire Lumber Company*. Madison: State Historical Society of Wisconsin, 1975.
Velte, Chas. H., *Historic Lake Poygan*. Neenah, WI 1976.
Vinette, Bruno, "Early Lumbering on the Chippewa." Madison: *Wisconsin Magazine of History*, June, 1926.
Wells, Robert W., *Daylight in the Swamp*. New York: Doubleday & Company, 1978.
Whitehouse, J.L., *A History of the Wolf River and It's Pioneers*, 1947-48-39. Typescript, Area Research Center, UW-Green Bay.
Williams, Roger, *The Loggers*. New York: Time-Life Books Inc., 1976.
Wyman, Walker D., *The Lumberjack Frontier*. Lincoln: University of Nebraska Press, 1969.

INDEX

S

T